Isaac Lea

Remarks on the Number of Unionidae

Isaac Lea

Remarks on the Number of Unionidae

ISBN/EAN: 9783337804091

Printed in Europe, USA, Canada, Australia, Japan

Cover: Foto ©berggeist007 / pixelio.de

More available books at **www.hansebooks.com**

Extracts from the Proceedings of the Academy of Natural Sciences of Philadelphia, of papers,
By ISAAC LEA.

[Communicated March 13th, 1860.]

Mr. Lea stated that when be made some remarks, a few weeks since, on the *Unionidæ* of the United States, he gave the number of them incorrectly by an inadvertence. He now desired to restate them numerically:

Unio, ...	465 species.
Margaritana, ...	26 "
Anodonta, ...	59 "
	550

To these may be added, new species in his cabinet not yet
  described, ... 30

  580

And to these may be added, for North America, known to
  inhabit Mexico, Honduras, Central America and
  one in Canada,   Unio,   29
                  Anodonta,   8
                      —   37

  617

It will be observed that we have not in North America either of the genera. *Triquetra,* (*Hyria,* Lam.,) *Prisodon,* (*Castalia,* Lam.,) *Monocondylœa, Mycetopus, Byssandonta,* or *Plagiodon.* They are all emphatically South American types, while there does not seem to inhabit the southern half of America a single species of *Margaritana,* (*Alasmodonta,* Say.) Ferussac has described a species (*A. incurva*) as coming from South America, but there is reasonable doubt of it. The *Monocondylœa* and *Margaritana* seem mutually to replace each other. The *Uniones* and *Anodontœ* prevail in both parts of the continent over all the other genera, both as to numbers and universality of distribution. The genus *Mulleria,* (*Acostea,* D'Orb.) has only been found in the tributaries of the Magdalena in New Granada.

Descriptions of Fifteen new Species of Uruguayan UNIONIDÆ

BY ISAAC LEA.

[Read March 13th, 1860.]

During the winter of 1858–59, R. B. Forbes, Esq., of Boston, whose name has been identified with so many works of philanthropy and public utility, organized an excursion to the La Plata, the Uruguay and Rio Negro rivers, in South America ; his object in part being to afford facilities for studying the natural history of the countries bordering on these waters. Professor J. Wyman, who accompanied him, has most kindly placed at my disposal all the specimens of the *Unionidæ* which he had been enabled to collect in these extensive southern fresh waters. In this very interesting collection I was surprised to find so many species which had not been before observed. These are now herein described, and consist of eleven *Uniones* and four *Anodontæ.* The whole number brought of these fresh water *Mulluscs*, was twenty-three species. Those heretofore described are *Prisodon truncatus,* Schum., (*Castalia ambigua,* Lam.,) *Unio Paranensis,* Lea., *U. parallelopipedon,* Lea., *Anodonta rotunda,* Spix, *A. trapezalis,* Lam., *A. lato-marginata,* Lea, *A. tenebricosa,* Lea, *A. Blainvilliana,* Lea. In addition there were three small species of *Cyrena,* two of which I have not ascertained, the third is the *variegata* of D'Orbigny. There was also a small species of *Cyclas.*

UNIO WYMANII.—Testâ lævi, anticè subsulcatâ, quadratâ, compressâ, ad latere planulatâ, inæquilaterali, posticè obtusè angulatâ, anticè rotundatâ ; valvulis subcrassis, anticè crassioribus ; natibus prominulis, ad apices divaricatè undulatis ; epidermide tenebroso-olivâ, vel cradintâ vel obsoletè radiatâ ; dentibus cardinalibus compressis, erectis, crenulatis, in utroque valvulo duplicibus ; lateralibus longis, crenulatis subcurvisque ; margaritâ argenteâ et valdè iridescente.

Hab.—Uruguay River, S. America.   Prof. J. Wyman.

UNIO URUGUAYENSIS.—Testâ lævi, anticè subsulcatâ, ellipticâ, inflatâ, subequilaterali, posticè obtusè angulatâ, anticè rotundatâ ; valvulis subcrassis, anticè crassioribus ; natibus subprominentibus, ad apices divaricatè undulatis ; epidermide virido-fuscâ, posticè tenebricosâ, politâ, obsoletè radiatâ ; dentibus cardinalibus compressis, crenulatis suberectisque ; lateralibus longis subrectisque ; margaritâ argenteâ et iridescente.

Hab.—Uruguay River, S. America.   Prof. J. Wyman.

UNIO PIGER.—Testâ lævi, ellipticâ, inflatâ, subequilaterali, posticè obtusè angulatâ, anticè obliquè rotundatâ ; valvulis crassiusculis, anticè paulisper crassioribus; natibus subprominentibus, inflatis, ad apices divaricatè undulatis ; epidermide nigro-fuscâ, striatâ, obsoletè radiatâ; dentibus cardinalibus compressis, crenulatis; lateralibus sublongis curvisque; margaritâ argenteâ ·t iridescente.

Hab.—Uruguay River, S. America. Prof. J. Wyman.

UNIO PERÆFORMIS.—Testâ lævi, subrotundâ, inflatâ, valdè inæquilaterali, posticè obtusè subangulatâ, anticè obliquè rotundatâ; valvulis subcrassis, anticè paulisper crassioribus; natibns vix prominentibus, inflatis; epidermide striatâ, nigro-virente, eradiatâ; dentibus cardinalibus parviusculis, compressis crenulatisque; lateralibus sublongis subrectisque; margaritâ argenteâ et iridescente.

Hab.—Uruguay River, S. America. Prof. J. Wyman.

UNIO NOCTURNIS.—Testâ lævi, subrotundâ, subcompressâ, inæquilaterali, anticè et posticè rotundatâ; valvulis crassis, anticè crassioribus: natibus prominulis, subinflatis; epidermide nigricante, anticè rugoso-striatâ, eradiatâ ; dentibus cardinalibus parviusculis, erectis; subcompressis, in utroque valvulo duplicibus; lateralibus sublongis valdè curvisque; margaritâ vel albâ vei salmonis colore tiuctâ.

Hab.—Uruguay River, S. America. Prof. J. Wyman.

UNIO FUNEBRALIS.—Testâ lævi, subrotundatâ, compressissimâ, inæquilaterali, anticè et posticè rotundatâ; valvulis crassis, anticè crassioribus; natibus prominulis, compressis; epidermide nigricante, striatâ, ad apices micante, eradiatâ ; dentibus cardinalibus parviusculis, subcompressis, tripartitis : lateralibus sublongis valdè curvisque ; margaritâ vel albâ vel salmonis colore tinctâ.

Hab.—Uruguay River, S. America. Prof. J. Wyman.

UNIO GRATUS.—Testâ lævi, subrotundâ, subinflatâ, inæquilaterali, anticè et posticè rotundatâ; valvulis subcrassis, anticè paulisper crassioribus; natibus subprominentibns, ad apices divaricatè undulatis; epidermide tenebroso-fuscâ, micantè, obsoletè radiatâ; dentibns cardinalibus parviusculis, compressis, striatisque; lateralibus sublongis subcurvisque : margaritâ albâ et iridescente.

Hab.—Uruguay River, S. America. Prof. J. Wyman.

UNIO DISCULUS.—Testâ lævi, snbrotundâ, valdè compressâ, valdè inæquilaterali, anticè et posticè rotundatâ ; valvulis crassiusculis, anticè paulisper crassioribus; natibus subprominentibus, ad apices paulisper divaricatè undulatis : epidermide tenebroso-castaneâ, minutè striatâ obsoletè radiatâque ; dentibus cardinalibns parviusculis, lamellatis crenulatisque; lateralibus sublongis, striatis cnrvisque ; margaritâ albâ et iridescente.

Hab.—Uruguay River, S. America. Prof. J. Wyman.

4

Unio piceus.—Testâ lævi, ellipticâ, subinflatâ, valdè inæquilaterali, posticè subrotundatâ, anticè obliquè rotundatâ; valvulis crassiusculis, anticè paulisper crassioribus; natibus prominulis; epidermide micante, nigrâ, striatâ obsoletè radiatâ vel eradiatâ; dentibus cardinalibus parviusculis, compressis, obliquis, in valvulo sinistro singulis; lateralibus sublongis subcurvisque; margaritâ cæruleâ albâ et iridescente.

Hab.—Uruguay River, S. America. Prof. J. Wyman.

Unio lepidus.—Testâ lævi, ellipticâ, subinflatâ, valdè inæquilaterali, posticè subrotuudatâ, anticè rotundâ; valvulis subtenuibus, anticè paulisper crassioribus; natibus prominulis, ad apices rugosè et divaricatè undulatis; epidermide politâ, fusco-virentè, striatâ, radiatâ; dentibus cardinalibus parviusculis, compressis, obliquis; lateralibus sublongis subcurvisque; margaritâ cæruleo-albâ -t valdè iridescente.

Hab.—Uruguay River, S. America. Prof. J. Wyman.

Unio Æthiops.—Testâ lævi, oblongâ, subinflatâ, ad laterè planulatâ, valdè inæquilaterali, posticè biangulatâ, anticè rotundatâ; valvulis crassiusculis, anticè crassioribus; natibus prominulis, planulatis, ad apices divaricatè undulatis; epidermide micante, nigrá, striatâ, eradiatâ; dentibus cardinalibus parviusculis, compressis, obliquis, suberectis crenulatisque; lateralibus prælongis, crenulatis rectisque; margaritâ albâ et iridescente.

Hab.—Uruguay River, S. America. Prof. J. Wyman.

Anodonta Wymanii.—Testâ lævi, ellipticâ, subinflatâ, inæquilaterali, posticè subbiangulatâ, anticè regulariter rotundatâ; valvulis crassis, anticè paulisper crassioribus; natibus prominulis, ad apices æquis; epidermide cinnomomeâ, vel eradiatâ vel obsoletè radiatâ; margaritâ roseâ et valdè iridescente.

Hab.—Uruguay River, S. America. Prof. J. Wyman.

Anodonta rubicunoa—Testâ alatâ, lævi, subrotundâ, inflatâ, subequilaterali, anticè et posticè rotundatâ; valvulis subtenuibus; natibus elevatis, tumidis, rosaceis; epidermide tenebroso-rufo-fuscâ, vel obsoletè radiatâ vel eradiatâ. margaritâ rufo-salmonis colore tinctâ et valdè iridescente.

Hab.—Uruguay River, S. America. Prof. J. Wyman.

Anodonta Forbesiana.—Testâ lævi, suboblongâ, ventricosâ, inæquilaterali. valvulis crassiusculis; natibus elevatis, inflatis; epidermide luteo-fuscâ. micante, vel eradiatâ vel obsoletè radiatâ; margaritâ albidâ et valdè iridescente.

Hab.--Uruguay River, S. America. Prof. J. Wyman.

ANODONTA URUGUAYENSIS.—Testâ lævi, obovatâ, ventricosâ, valdè inæquilaterali: valvulis subcrassis, anticè paulisper crassioribus; natibus subelevatis. tumidis; epidermide tenebroso-olivâ, eradiatâ; margaritâ cæruleo-albâ et valdè iridescente.

Hab.—Uruguay River, S. America. Prof. J. Wyman.

Descriptions of Four New Species of UNIONIDÆ from Brazil and Buenos Ayres.

BY ISAAC LEA.

(Read March 13th, 1860.)

UNIO TRIFIDUS.—Testâ lævi, obliquo-oblongâ, ad laterè planulatâ, valdè inæquilaterali, posticè acutè augulatâ, anticè rotundâ; valvulis crassiusculis. anticè crassioribus; natibus prominentibus, ad apices rugosè et divaricatè undulatis; epidermide micante, luteo-viridi, eradiatâ; dentibus cardinalibus grandibus, trifidis, sulcatis; lateralibus longis, crenulatis, in valvulo dextro trifidis; margaritâ argenteâ et iridescente.

Hab.—Buenos Ayres, South America. M. D'Orbigny.

UNIO PATELLOIDES.—Testâ lævi, subrotundâ, subcompressâ, subæquilaterali. anticè et posticè rotundatâ; valvulis subcrassis, anticè crassioribus; natibus prominulis, ad apices divaricatè undulatis; epidermide tenebroso-castaneâ. striatâ, eradiatâ; dentibus cardinalibus longis, compressis, obliquis, crenulatis corrugatisque; lateralibus longis, crenulatis curvisque; margaritâ argenteâ et iridescente.

Hab.—Amazon River, Brazil. Captain George Brown. Rio Plata. H. Cuming.

ANODONTA AMAZONENSIS.—Testâ lævi, transversâ, subinflatâ, valdè inæquilaterali, posticè subbiangulatâ, anticè rotundâ; valvulis subcrassis; natibus subelevatis, tumidis; epidermide micante, tenebroso-viridi, nigricante, vel eradiatâ vel obsoletè radiatâ; margaritâ intus subroseâ et valdè iridescente.

Hab.—Upper Amazon, Brazil. C. M. Wheatley.

ANODONTA MORICANDII.—Testâ lævi, obliquè quadratâ, subinflatâ, ad laterè planulatâ, valdè inæquilaterali, posticè obtusè angulatâ et biante; anticè obliquè rotundatâ et valdè biante; valvulis tenuibus, diaphinis; natibus subprominentibus; epidermide luteo-olivâ. politâ, obsoletè radiatâ, margaritâ cæruleo-albâ et valdè iridescente.

Hab.—Bahia, Brazil. S. Moricand, Geneva.

[Communicated March 20th, 1860.]

Mr. Lea read extracts from letters of Dr. Lewis, of Mohawk, New York, on the subject of the coloring matter of the nacre of the genus *Unio*, and exhibited some fine specimens to illustrate the subject. The following extracts will fully convey Dr. Lewis's ideas on this subject which has much interest with the naturalist.

"I hinted something about *Uniones* being colored with an oxide or salt of gold. My reasons for this are derived from observing some singular phenomena in colors on submitting shells to the action of chloride of gold, and then bringing them in contact with tin. Whether a stannate of gold formed and precipitated on the shells or not, I cannot say, but the colors were very much intensified. It is to be remarked that the colors of such shells as *Unio complanatus* and of *U. ligamentinus*, when colored, are such as result from the presence of gold in a state of atomic division and dissemination in a semi-opake body. I think nitro-muriatic acid with a minute trace of gold in it, if applied to shells, will produce *colors*, but I never have satisfactorily demonstrated this. My observations are derived from having once used acid in which was a small quantity of gold, too small to be reclaimed."

"I notice that colors are most brilliant in regions where gold may be suspected. In the Lake regions of the Western States, minerals are abundant, and the conditions are not incompatible with the supposition that gold is sparingly disseminated among them, in quantities too small perhaps to be available, *but no doubt it is there.*"

"As regards *colors* in the *nacre* of *Uniones*, you are correct in saying that Uniones *are colored where there is no gold.* But there are some species that are *not* colored unless you find them in some particular localities. If that is taken into consideration we shall, perhaps, be more ready to accept the *gold* theory. Modern investigations show that gold exists in soils that, until they were rigidly tested, were not suspected to contain it. In fact I am disposed to believe that gold is more universally disseminated than is generally supposed."

"But, the question is one I take no particular interest in, except that it presents itself incidentally. I know one fact that you also know. That of two streams producing identically the same species, one will give a large proportion of white nacres, and the other will present *colored* nacres, and usually we also notice another phenomenon—*a greater brilliancy* of nacre where rich colors abound. In this case I have my *private opinion* that gold produces its peculiar *tonic* effect, for *tonic* it is under certain circumstances by increasing the secretions."

"To have gold in a shell, it is not necessary it should be an oxide. It is only necessary it should have been received into the circulation of the animal, in solution as *chloride*, or some other possible soluble form that chemistry has not brought to light; and when once in the circulation it may be eliminated by being deprived of its solving principle and excreted or secreted with the other solid matter that enters into the formation of the shell. The stannate of gold, or purple of Cassius, may be wholly deprived of the tin associated with it, yet retain its purple color, and its condition of atomic division, if so you are pleased to call it. But I only offer this as suggestive of something for those interested to follow further. I am not enough of a chemist to develop any facts out of a suspicion of this kind."

Mr. Lea remarked, after reading the above extracts, that the *purple, pink and salmon* color of many of our American *Unionidæ* had had his attention from the period of his first studying this beautiful and interesting family, more than thirty years since. Without having experimented himself upon them, he was aware that no chemist had been able to detect the presence of a metal or other elementary body. He therefore thought it likely to be caused by the presence of some organic body which had not yet been detected; such is supposed by chemists to be the case with the colored fluates of lime, colored quartz, &c. What Dr. Lewis states as regards the colors being more frequent and more intense in the waters of Michigan and in the streams leading into the northern great lakes from the southern side, is very true. The *Unio rectus* is usually white in the Ohio, though sometimes tinted with purple and salmon color, while in the more northern waters it is usually of a fine rich purple or salmon. Two specimens from the upper Mississippi, brought by Dr. Cooper, were exhibited by Mr. Lea, which were of exquisite purple and salmon. The *Unio ligamentinus* has probably never been found pink or purple in the Ohio, while at Grand Rapids, Michigan, those with a fine pink and salmon color are very common. The *Margaritana margaritifera* of Columbia river and its tributaries has a fine purple nacre in almost all the specimens, rarely white, while those in the rivers of Pennsylvania, Connecticut and Massachusetts are almost universally white, as those from the northern part of Europe are also.

Dr. Draper had informed Mr. Lea that he had calcined some of these purple shells, but that they had burned white and he had not detected any metallic substance in their composition. The subject was certainly one well worth the pursuit, as no doubt could remain that the color was derived from some foreign substance entering into the composition of some individuals, while others were free from it. It was not an uncommon case to find the dorsal portion of the nacre to be pink or purple while the other portions were white, and this was also sometimes the case with the cavity of the beaks. Mr. Lea did not believe the color arose, as some persons supposed, from the structure of the surface of the nacre dividing the rays of light by thin laminations. This division of color was exhibited in almost every species, and is what naturalists call the "pearly hue," oftentimes of great beauty, but quite a different matter from the pink, purple and salmon color of the mass of the carbonate of lime composing the substance of the valves.

Descriptions of Four New Species of MELANIDÆ of the United States.

BY ISAAC LEA.

(Read March 20th, 1860.)

SCHIZOCHILUS SHOWALTERII.—Testâ transversè costatâ, subcylindraceâ, crassâ, castaneâ, minutè striatâ ; spirâ elevatâ; suturis impressis; anfractibus subplanulatis ; fissurâ submagnâ, profurdâ; aperturâ subparvâ, ellipticâ, intus vittatâ ; columellâ subcrassâ; labro paulisper crenulato.

Hab.—Coosa river, Uniontown, Alabama.   E. R. Showalter, M. D.

ANCULOSA SHOWALTERII.—Testâ valdè costatâ, suborbiculari, crassâ, tenebroso-fuscâ, nigricante, exilissimè striatâ; spirâ brevissimâ ; suturis valdè impressis ; anfractibus inflatis, septenis transversis costis indutis ; aperturâ magnâ, sub-rotundâ, supernè subangulatâ, internè tenebroso-vittatâ ; columellâ crassâ. planulatâ, tenebroso-fuscâ ; labro valdè extenso et valdè crenulato.

Hab.—Coosa river, Uniontown, Alabama.   E. R. Showalter, M. D.

MELANIA CRENATELLA.—Testâ transversè striatâ, turrito-subulatâ, subcostatâ, paulisper plicatâ, subtenui, tenebroso-fuscâ, nigricante ; spirâ elevatâ, ad apices crebrè plicatâ ; suturis valdè impressis ; anfractibus septenis, planulatis, trans-versis costis indutis ; aperturâ parvâ, ellipticâ, intus vittatâ ; columellâ albidâ, incurvatâ ; labro subcontracto et valdè crenulato.

Hab.—Coosa river, Uniontown, Alabama.   E. R. Showalter, M. D.

MELANIA NEWBERRYI.—Testâ lævi, ovato-conicâ, subtenui, tenebroso-fuscâ, trivittatâ, infernè suturis luteâ; spirâ subelevatâ ; suturis valdè impressis ; anfractibus senis, inflatis; aperturâ parviusculâ, ovato-rotundatâ, intus albidâ et vittatâ ; columellâ albidâ, incurvatâ ; labro inflato.

Hab.—Upper des Chutes river, Oregon Territory.   J. S. Newberry, M. D.

Descriptions of Five New Species of UNIONES from North Alabama.

BY ISAAC LEA.

[Read March 20th, 1860.]

Unio podicus.—Testâ lævi, subtrigonâ, compressâ, inæquilaterali, posticè obtusè angulatâ, anticè rotundâ ; valvulis subcrassis, anticè crassioribus ; natibus prominulis, ad apices rugoso-undulatis; epidermide luteo-fuscâ, micante, virido-radiatâ ; dentibus cardinalibus crassiusculis, erectis, compressis crenulatisque ; lateralibus subcurtis, crassis subcurvisque ; margaritâ albâ et iridescente.

Hab.—North Alabama, Prof. Tuomey ; and Florence, Alabama, L. B. Thornton, Esq.

Unio camelopardilis.—Testâ lævi, oblongâ, subinflatâ, inæquilaterali, posticè obtusè biangulatâ, anticè regulariter rotundatâ; valvulis subtenuibus, anticè crassioribus; natibus prominulis, ad apices rugoso-undulatis ; epidermide luteâ, politâ, undiquè virido-maculatâ; dentibus cardinalibus parvis, erectis, compresso-pyramidatis crenulatisque ; lateralibus longis, lamellatis subrectisque ; margaritâ luteo-albâ et valdè iridescente.

Hab.—North Alabama, Prof. Tuomey.

Unio fucatus.—Testâ lævi, ellipticâ, subinflatâ, valdè inæquilaterali, posticè subbiangulatâ, anticè rotundatâ; valvulis tenuibus, anticè paulisper crassioribus ; natibus prominulis, ad apices undulatis; epidermide olivo-luteâ, micante, undiquè virido-maculatâ ; dentibus cardinalibus parvis, compresso-conicis, crenulatis, in utroque valvulo duplicibus ; lateralibus longis, lamellatis subcurvisque ; margaritâ vel cæruleâ vel luteo-albâ et valdè iridescente.

Hab.—North Alabama, Prof. Tuomey. Tuscumbia, L. B. Thornton, Esq.

Unio discrepans.—Testâ lævi, ellipticâ, subinflatâ, ad laterè subplanulatâ, valdè inæquilaterali, posticè obtusè biangulatâ, anticè rotundatâ ; valvulis subtenuibus, anticè crassioribus ; natibus prominulis ; epidermide luteo-olivâ, micante, radiatâ ; dentibus cardinalibus parvis, compresso-conicis crenulatisque ; lateralibus longis, lamellatis subcurvisque ; margaritâ vel albâ vel purpureâ et valdè iridescente.

Hab.—North Alabama, Prof. Tuomey.

2

¹Unio ᴘʟᴀɴɪᴄᴏsᴛᴀᴛᴜs.—Testá lævi, ellipticá, compressá, ad latere subplanulatá, valdè inæquilaterali, posticè obtusè biangulatá; anticè rotundatá; valvulie tenuibus, diaphanis, anticè paulisper crassioribus; natibus prominulis, ad apices undulatis; epidermide olivaceá, undiquè radiatá; dentibus cardinalibus parvis, coniois, crenulatis, in utroque valvulo duplicibus; lateralibus longis lamellatis subcurvisque; margaritá vel cæruleo-albá vel purpurascente et valdè iridescente.

Hab.—Tuscumbia, Alabama, L. B. Thornton, Esq.

Unio sᴄɪᴛᴜʟᴜs.—Testâ lævi, ellipticá, inflatá, valdè inæquilaterali, posticè obtusè biangulatâ, anticè rotundatâ; valvulis subtenuibus, anticè crassioribus; natibus prominentibns, ad apices undulatis; epidermide luteâ, undique viridoradiatâ; dentibus cardinalibus parviusculis, erectis, acuminatis, crenulatis, in utroque valvulo duplicibus; lateralibus longis, lamellatis subrectisque; margaritâ albâ et valdè iridescente.

Hab.—Tuscumbia, Alabama, L. B. Thornton, Esq.

[Communicated April 10th, 1860.]

Mr. Lea remarked that he had recently received from Prof. J. Wyman specimens in alcohol of two species of *Anodonta* from the Uruguay River, South America, descriptions of the soft parts of which he had made, and intended, at a future time, to publish in the Journal at length; but he wished at present to mention that he had found a form of *Palpi* (mouth lips) different from any of the *Unionidæ* which had come under his notice from any other part of the world. The form of the *Palpi* heretofore described have always been obliquely or transversely elliptical or subtriangular, while these two species, *An. Wymanii*, Lea, and *An. lato-marginata*, Lea, are round, and the pair on either side only joined above, the edges being entirely free. It is greatly to be regretted that more or all the South American *Unionidæ* could not have been examined, as regards their soft parts, to ascertain if this difference of form of the *Palpi* should be persistently different in all the South American *Unionidæ*, or only with this member of the family—the *Anodontæ*.

[Communicated May 8th, 1860.]

Mr. Lea mentioned that he had recently received a letter from Dr. Showalter of Uniontown, Alabama, in which he mentions that specimens of *Physa (gyrina)* Say, which he sent on, were obtained in an open neglected cistern, and in a trough of water supplied by an Artesian well ten miles from the town. Dr. S. expressed his surprise that these *Physæ* should find their homes so soon at these Artesian wells. There are no streams or pools near to these wells, but in a few years after they are bored and water supplied, these shells may with certainty be found. Mr. Lea went on to mention that he had nearly 30 years ago found an undescribed species of *Lymnæa*, accompained by *Physa heterostropha*, Say, in a small artificial pond on the high grouuds near to the Falls of Schuylkill, about four miles north of Market Street, now within the limits of this City. He published an acconnt of it in April 1834, in the Trans. Am. Phil. Soc. under the name of *acuta*. The pond was small and dug out for $1\frac{1}{2}$ to 2 feet deep, simply for the supply of rain water for cattle. Afterwards it dried up and the shells were no longer to be obtained there. He never found this *Lymnæc* in, any other habitat; but many years snbsequently, Dr. Ingalls, of Greenwich, N. Y., near to Lake Champlain, sent him several specimens of what he regard-ed as a new *Lymnæa*, but which was at once recognised as the *acuta*, heretofore found only in the one habitat near the Falls of Schuylkill. In the minds of some zoologists a difficulty exists as to existence of species in such constricted, isolated points as mentioned above, but that difficulty in Mr. Lea's mind was done away with under the belief that very young molluscs may be transported on the feet of birds from distant points, or on those of cattle going to drink from one place to another. The idea of spontaneous generation could not of course be for one moment admitted.

[Communicated May 22d, 1860.]

Mr. Lea called the attention of the members to two very remarkable specimens of *Echinus*, perforating rocks, which he had recently received from Mr. Cailliaud, of Nantes, the Egyptian traveller. He also exhibited a specimen of Sandstone from Payta in Peru, which contained *Petricola, Lithophagus*, &c. He reminded the members that he had presented to the Academy a very remarkable specimen, which he had received about two years since from Mr. Cailliaud, being a mass of *gneiss* which had been perforated by *Pholades*. When Mr. Cailliaud, who had advocated, contrary to the opinion of most naturalists, the theory that some of the Molluscs bored the rocks by friction and not by decomposition, found that *gneiss* and *granite* and other *silicious rocks* were penetrated by them, he entirely settled that question, for there are no acids known which will decompose silex. Mr. Lea remarked that the two specimens now on the table were still more remarkable. The smaller one consisted of two specimens of *Echinus lividus*, Lam., which had buried themselves in the solid *granite*, one of them having made a circular hole $1\frac{1}{2}$ inch deep, and 2 inches wide. This specimen came from the granite coast of the Loire-Inférieure The second specimen consisted of quite a congress of individuals of the same species, imbedded in a solid mass of hard *Silurian Sandstone*, from the Bay of Douarneuez, in the Department of Finistère. In this beautiful specimen there are five individuals nestled in their circular holes, worked out in this hard stone by the attrition of their teeth, and there are three holes vacated. The specimen is 5 inches by $6\frac{1}{2}$, and there being eight holes in all, their circumferences nearly impinge on each other. Mr. Cailliaud is entirely satisfied that the boring is purely mechanical, that the five teeth are the instruments of exploitation, and that it is by the percussion of their points on the rocks that these holes are effected. The teeth are in form like the rodents, and constantly increase as worn at the outer extremity. He made a hole five millimetres deep and forty in circumference with a bundle of the teeth in an hour. One of the colonies which he examined was in a bay, and contained about two thousand holes, each one filled, and at low water they were but a short distance below the surface. Some of the specimens were not larger than a pea, and probably only five days old. The holes were not all made by the present occupants, most of them probably being very old and having successive inhabitants. Mr. Cailliaud mentioned in his letter to Mr. Lea that he shortly expected to receive from Guadaloupe an *oval Echinus* which had made its *oval hole* in the mass of *Madreporite*.

Descriptions of Fourteen new species of Schizostomæ, Anculosæ and Lithasiæ.

BY ISAAC LEA.

[Read May 22d, 1860.]

It will be observed that I have in this paper adopted my first name (*Schizostoma*) for the division of those *Melanidæ* which have a cut or fissure in the upper portion of the last whorl. This name I proposed in December, 1842. Subsequently finding that it was used by Bronn in 1835 I abandoned it, and proposed the name of *Schizochilus* as a substitute, (Obs. on the Genus Unio. v. 5. p. 51, 1852.) I am now satisfied that Bronn's name was applied to the same genus—*Euomphalus*—which Sowerby established in 1814, (Min. Conch. tab. 45.) This evidently liberates my original name, and Herrmannsen, in the Appendix to his "Generum Malacozorum," very properly restores it. It was supposed that this was the *Melatoma* of Swainson, and Mr. Anthony adopted this name. But it is evident that Mr. Swainson's *Melatoma* is not my *Schizostoma*. By reference to his figure (Malacology, p. 342, f. 104) it will be observed at once that there has never been observed in the United States any of the group of which that figure is the type, while it is known that they exist in the islands of the Indian Ocean. Mr. Swainson says (p. 202) that his *Melatoma* was "founded upon a remarkable Ohio shell" sent by Rafinesque. Now, as no member of the family *Melanidæ* with a cut in the lip has ever been found in the Ohio, where such hosts of active collectors have since pursued their investigations, it is perhaps beyond the bounds of possibility that the specimen sent by Rafinesque so eminently careless and reckless as he always was, should ever have been found there. Indeed, if the specimen figured was sent by Mr. Rafinesque to Mr. Swainson, then the question would arise whether it had not been obtained by Mr. R. from some dealer or collector, who may have obtained it from Asia. I have no doubt of the *Melatoma costata*, which Mr. Swainson has figured, being exotic, and belonging to a group probably from the Philippine Islands. Mr. Anthony says, page 64, Proc. A. N. S. 1860, that "it may be doubted whether Mr. Lea's first name will not eventually prevail, since, before he published *Schizostoma*, Bronn's genus of the same name had been called a synonym of *Bifrontia*, Desh." And that "H. and A. Adams (Gen. Rec. Moll. 1, 105) do not appear correct in giving preference to *Gyrotoma* over *Schizostoma*, Lea," &c. Notwithstanding this, Mr. Anthony in this paper, where he describes nine supposed new species of this genus, adopts the generic name of *Gyrotoma*. It may be added here that Dr. Gray, in his *Genera of Recent Mollusca*, gives *Melatoma* to

Mr. Antbony, not to Swainson, while he does not notice the name of *Schizos-toma.* Mr. A. does not pretend to claim it, of course, but adopts *Gyrotoma,* Mr. Shuttleworth's name, proposed in 1845, which being three years later cannot have precedence.

The genus *Schizostoma* seems to be capable of being divided into two natural groups in the form of the *fissura,* the cut in the lip. In one group this fissura is deep and direct, that is parallel with the suture or upper edge of the whorl; in the other it is not deep and is oblique to the suture. In Mr. Anthony's paper (Proc. Acad. Nat. Sci. Feb., 1860) I recognize several of my old species. His *Gyrotoma demissa* I believe to be my *Schizostoma con-stricta.* His *G. quadrata* to be my *S. incisa.*

SCHIZOSTOMA CASTANEA.—Testâ carinatâ, conicâ, subcrassâ, tenebroso-fuscâ. imperforatâ; spirâ elevatâ; suturis valdè impressis; anfractibus senis, planu-latis, unicarinatis, quadrivittatis; fissurâ rectâ, angustâ profundâque; aperturâ parviusculâ, ellipticâ, intus vittatâ, ad basim subrotundatâ; columellâ albâ, in-crassatâ; labro acuto, vix sinuato.

Hab.—Coosa River, Alab. E. R. Showalter, M. D.

SCHIZOSTOMA GLANS.—Testâ lævi, ovato-conicâ, inflatâ, subcrassâ, luteo-cor-neâ, striatâ, imperforatâ; spirâ obtusè elevatâ: suturis regulariter impressis; anfractibus senis, obsoletè vittatis, ultimo subgrandi; fissurâ rectâ, angustâ profundâque; aperturâ parviusculâ, ellipticâ, intus albidâ, ad basim obtusè angulatâ; columellâ albidâ, supernè incrassatâ; labro-acuto, subsinuato.

Hab.—Coosa River, Alab. E. R. Showalter, M. D.

SCHIZOSTOMA GLOBOSA.—Testâ transversè striatâ, globosâ, subtenui, luteola, imperforatâ; spirâ curtâ, obtusè conoideâ; suturis impressis; anfractibus qua-ternis, trivittatis, ultimo grandi; fissurâ rectâ, angustâ brevique; aperturâ subgrandi, ellipticâ, intus vittatâ, ad basim angulatâ; columellâ albâ. incur-vatâ; labro acuto, expanso.

Hab.—Alabama. E. R. Sbowalter, M. D.

SCHIZOSTOMA VIRENS.—Testâ subnodulosâ, curtâ, inflatâ, subcrassâ, tenebroso-viridi, exilissimè striatâ, imperforatâ; spirâ obtusâ; suturis impressis; anfrac-tibus subplanulatis et trivittatis; fissurâ obliquâ brevique; aperturâ elongatâ. subpyriformi, intus tenebroso-vittatâ; columellâ supernè purpuratâ et incras-satâ; labro acuto, sinuato.

Hab.—Coosa River, Alab. E. R. Showalter, M. D.

SCHIZOSTOMA GLANDULA.—Testâ lævi, curtâ, inflatâ, subcrassâ, luteo-corneâ. exilissimè striatâ, imperforatâ: spirâ obtusâ: suturis valdè impressis; anfractibus senis, vittatis, ultimo magno et tumido ; fissurâ obliquâ brevique ; apertura subgrandi, ellipticâ, intus albidâ ; columellâ albidâ, supernè incrassatâ ; labro acuto, subsinuato.

Hab.—Coosa River, Alab. E. R. Sbowalter, M. D.

SCHIZOSTOMA WETUMPKAENSIS.—Testâ striatâ, ovato-cylindraceâ, crassâ, pallido-fuscâ, perforatâ ; spirâ obtusâ, conoideâ ; suturis valdè impressis ; anfractibus senis, vittatis, planulatis, ultimo grandi ; fissurâ obliquâ brevique ; aperturâ grandi, ovatâ, intus vittatâ, ad basim obtusè angulatâ ; columellâ albâ, supernè incrassatâ ; labro acuto, sinuato.

Hab.—Coosa River, at Wetumpka, Alabama. E. R. Sbowalter, M. D.

SCHIZOSTOMA ALABAMENSIS.—Testâ striatâ, ellipticâ, robustâ, luteo-olivaceâ, imperforatâ, spirâ obtuso-conoideâ ; suturis valdè impressis ; anfractibus senis, vittatis, subinflatis, ultimo pergrandi ; fissurâ obliquâ subbrevique ; aperturâ subgrandi, ovatâ, intus vittatâ, ad basim rotundatâ ; columellâ albâ, infernè et supernè paulisper incrassatâ ; labro acuto, sinuato.

Hab.—Alabama. B. W. Budd, M. D., and E. R. Sbowalter, M. D.

SCHIZOSTOMA HARTMANII.—Testâ lævi, subcylindraceâ, crassâ, luteo-corneâ, imperforatâ ; spirâ elevatâ ; suturis valdè impressis ; anfractibus planulatis, ultimo subgrandi ; fissurâ rectâ subbrevique ; aperturâ parviusculâ, ovatâ, intus albâ, ad basim obtusè angulatâ ; columellâ albâ, incurvâ, infernè paulisper incrassatâ ; labro acuto, sinuato.

Hab.—Coosa River, Alab. W. D. Hartman, M. D.

SCHIZOSTOMA PUMILA.—Testâ striatâ, turbonatâ, subtenui, pallido-corneâ, imperforatâ ; spirâ valdè obtusâ ; suturis valdè impressis ; anfractibus senis, ventricosis, ultimo permagno ; fissurâ rectâ subbrevique ; aperturâ parviusculâ, ovatâ, intus albâ, ad basim angulatâ et subcanaliculatâ ; columellâ albâ, contortâ, infernè incrassatâ ; labro acuto, sinuato.

Hab.—Alabama. B. W. Budd, M. D.

ANCULOSA FORMOSA.—Testâ lævi, globosâ, subtenui, diaphanâ, vel luteolâ vel crocatâ, valdè vittatâ et maculatâ ; spirâ depressâ vix conspicuâ ; suturis impressis ; anfractibus ternis, ultimo magno et valdè ventricoso ; aperturâ grandi, subrotundâ, intus pallido-crocatâ et tenebroso-vittatâ ; columellâ infernè et supernè incrassatâ et pallido-purpuratâ ; labro acuto et valdè expanso.

Hab.—Coosa River, Shelby Co., Alabama. E. R. Sbowalter, M. D.

ANCULOSA CONTORTA.—Testâ lævi, globoso-ovoideâ, crassâ, luteo-corneâ ; spirâ elevatâ ; suturis valdè impressis ; anfractibus inflatis, obsoletè transversè

striatis; aperturâ parvâ, subrotundâ, contractâ, intus luteo-albâ; columellâ incrassatâ; labro acuto, expanso.

Hab.—Coosa River, at Watumpka, Alab. E. R. Showalter, M. D.

ANCULOSA VITTATA.—Testâ lævi, subglobosâ, crassâ, luteolâ, valdé vittatâ;
spirâ obtusâ; suturis impressis; anfractibus quarternis, inflatis, ultimo magno
et ventricoso; aperturâ rotundâ, in faucibus valdé constrictâ, intus vittatâ;
columellâ valdé incrassatâ, planulatâ, purpuratâ; labro acuto, expanso.

Hab.—Coosa River, at Watumpka, Alabama. E. R. Showalter, M. D.

LITHASIA SHOWALTERII.—Testâ lævi, ovato-cylindraceâ, subcrassâ, luteo-corneâ, vittatâ; spirâ obtusé conoideâ; suturis valdé impressis, anfractibus senis,
ultimo magno et planulato: aperturâ grandi, subovatâ, elongatâ, intus albidâ,
tenebroso-vittatâ, ad basim obtusé angulatâ; columellâ infernè et supernè incrassatâ, incurvâ; labro acuto et subconstricto.

Hab.—Coosa River, at Watumpka, Alabama. E. R. Showalter, M. D.

LITHASIA NUCLEA.—Testâ lævi, ellipticâ, luteo-olivâ, crassâ, solidâ, trivittatâ;
spirâ obtusé conoideâ; suturis impressis; anfractibus quinis, ultimo magno et
paulisper inflato; aperturâ parviusculâ, ovato-rotundâ, intus albidâ, trivittatâ,
ad basim recurvatâ; columellâ infernè et supernè incrassatâ, incurvâ; labro
acuto.

Hab.--Coosa River, Alabama. E. R. Showalter, M. D.

Descriptions of Two New Species of UNIONES from Georgia.

BY ISAAC LEA.

[Read July 3d, 1860.]

Unio linguæformis.—Testâ lævi, ellipticâ, compressâ, ad laterè planulatâ, inæquilaterali, posticè obtusè biangulatâ, anticè rotundatâ, valvulis subtenuibus, anticè crassioribus; natibus prominulis; epidermide pallido-luteâ, subnitidâ, viridi-radiatâ, dentibus cardinalibus parviusculis, obtuso-conicis, crenulatis, in utroque valvulo duplicibus; lateralibus sublongis, lamcllatis subcurvisque; margaritâ albâ et valdè iridescente.

Hab.—Columbus, Georgia.

Unio dispar.—Testâ lævi, ellipticâ, subinflatâ, ad laterè paulisper planulatâ, valdè inæquilaterali, posticè obtusè angulatâ, anticè rotundatâ; valvulis subcrassis: natibus prominulis, ad apices undulatis; epidermide vel luteâ vel olivâ et valdè radiatâ; dentibus cardinalibus parviusculis, compressis, in utroque valvulo duplicibus, erectis, crenulatis; lateralibus longis, lamellatis curvisque, margaritâ albâ et iridescente.

Hab.—Columbus, Georgia, Bishop Elliott and G. Hallenbeck.

Descriptions of Three New Species of UNIONES from Mexico.

BY ISAAC LEA.

[Read July 3d, 1860.]

Unio Cocchianus.—Testâ lævi, quadratâ, inflatâ, biemarginatâ, ad laterè sulcatâ, inæquilaterali, posticè biangulatâ, anticè rotundâ; valvulis crassis, anticè crassioribus; natibus prominentibus: epidermide olivo-fuscâ, striatâ, eradiatâ: dentibus cardinalibus subgrandibus, erectis, rugoso-striatis, crenulatis: lateralibus sublongis, crassis curvisque; margaritâ argenteâ et valdè iridescente.

Hab.—Rio Salado, New Leon, Mexico, L. Berlandier, M. D.

Unio Salaooensis.—Testâ lævi, obovatâ, inflatâ, inæquilaterali, posticè et anticè rotundatâ; valvulis subtenuibus, anticè paulisper crassioribus; natibus prominulis, lævibus; epidermide luteo-olivâ, politâ radiatâque; dentibus cardinalibus parvis, lamellatis, obliquis; lateralibus sublongis, lamellatis subcurvisque; margaritâ cæruleo-albâ et iridescente.

Hab.—Rio Salado, New Leon, Mexico, L. Berlandier, M. D.

3

UNIO COGNATUS.—Testâ lævi, ellipticâ, crassâ, subinflatâ, valdè inæquilaterali, posticè obtusè angulatâ, anticè rotundatâ ; valvulis crassis, anticè crassioribns ; natibus subprominentibus ; epidermide luteolâ radiatâque ; dentibus cardinalibus grandibus, crassis, pyramidatis, rugoso-striatis ; lateralibus subcurtis, subcurvis crassisque ; margaritâ albâ et valdè iridescente.

Hab.—Rio Salado, New Leon, Mexico, L. Berlandier, M. D.

Descriptions of Seven New Species of UNIONIDÆ from the United States.

BY ISAAC LEA.

[Read July 3d, 1860.]

UNIO LESLEYI.—Testâ lævi, obliquâ, subcompressâ, ad laterè planulatâ, valdè inæquilaterali, posticè angulatâ, anticè rotundâ ; valvulis crassis, anticè crassioribus ; natibus subelevatis ; epidermide luteolâ, radiis interruptis ; dentibus cardinalibus crassis, compresso-conicis, obliquis ; lateralibus longis, crassis subcurvisque ; margaritâ albâ et iridescente.

Hab.—Kentucky, Mr. Joseph Lesley. Tennessee, Mr. J. G. Anthony.

UNIO CASTUS.—Testâ lævi, inflatâ, inæquilaterali, posticè angulatâ, anticè rotundâ, valvulis crassiusculis, anticè crassioribus ; natibus subprominentibus ; epidermide micante, luteo-viridi, radiatâ ; dentibus cardinalibus subgrandibus, compresso-conicis, striatis crenulatisque ; lateralibus subbrevibus, rectis, lamellatis striatisque ; margaritâ albâ et iridcscente.

Hab.—South Carolina, Prof. Tuomey.

UNIO LINDSLEYI.—Testâ lævi, ellipticâ, compressâ, ad laterè planulatâ, valdè inæquilaterali, posticè subbiangulatâ, anticè obliquè rotundatâ ; valvulis subcrassis, anticè crassioribus ; natibus prominulis ; epidermide vel luteâ vel luteo-olivâ, micante, undiquè virido-maculatâ ; dentibus cardinalibus parviusculis, compresso-conicis crenulatisque ; lateralibus longis, crassis cnrvisque ; margaritâ albâ et iridescente.

Hab.—Tennessee, President J. B. Lindsley.

UNIO PERPICTUS.—Testâ lævi, ellipticâ, subinflatâ, valdè inæquilaterali, posticè obtusè biangulatâ, anticè rotundatâ ; valvulis tenuibus, diaphanis, anticè crassioribus ; natibus prominulis, ad apices undulatis ; epidermide luteo-olivâ, subnitidâ, undiquè virido-radiatâ ; dentibus cardinalibus parvis, erectis, conicis, crenulatis, in utroque valvulo duplicibus ; lateralibus longis, lamellatis rectisque ; margaritâ cærnleo-albâ et valdè iridescente.

Hab.—Bull River, Tennessee, President Estabrook, and Holston River, Prof. Troost.

UNIO EIGHTSII.—Testâ plicatâ, quadratâ, subcompressâ, maximè undulatâ usque ad natium apices, valdè inæquilaterali; valvulis crassissimis, anticè crassioribus; natibus elevatis, compressis, ad apices rugoso-undulatis; epidermide tenebroso-fuscâ, eradiatâ, striatâ; dentibus cardinalibus magnis, crassis et valdè striatis; lateralibus longis, crassis, lamellatis subcurvisque; margaritâ albâ et valdè iridescente.

Hab.—Texas and Sabinas River, New Leon, Mexico, James Eights, M. D.

UNIO QUADRANS.—Testâ lævi, quadratâ, valdè ventricosâ, subæquilaterali. posticè obtusè angulatâ, anticè subtruncatâ; valvulis crassis, anticè paulisper crassioribus; natibus elevatis, tumidis; epidermide tenebroso-fuscâ, eradiatâ, transversè striatâ; dentibus cardinalibus magnis, valdè compressis, striatis crenulatisque; lateralibus longis, crassis curvisque; margaritâ albâ et iridescente.

Hab.—Texas, Mr. C. M. Wbeatley.

ANODONTA KENNERLYI.—Testâ lævi, ellipticâ, subcylindraceâ, subventricosâ. valdè inæquilaterali, anticè subtruncatâ, posticè biangulatâ; valvulis tenuissimis, diapbanis; natibus vix prominentibus, ad apices exilissimè undulatis; epidermide luteo-olivâ, micante, ad margine striatâ, eradiatâ: margaritâ cæruleo-albâ et valdè iridescente.

Hab.—Chiloweyucb Depot, near Puget's Sound, Washington Territory, C. B. Kennerly, M. D.

Descriptions of Six New Species of UNIONIDÆ from Alabama.

BY ISAAC LEA.

[Read July 3d, 1860.]

UNIO SHOWALTERII.—Testâ lævi, subrotundâ, crassâ, sublenticulari, æquilaterali; valvulis crassis, anticè crassioribus; natibus elevatis, tumidis; epidermide tenebroso-fuscâ, eradiatâ; dentibus cardinalibus crassis, erectis, compressis, in utroque valvulo duplicibus; lateralibus brevibus, percrassis corrugatisque; margaritâ argenteâ et valdè iridescente.

Hab.—Coosa River, Watumpka, Alab., E. R. Showalter, M. D.

UNIO HARTMANIANUS.—Testâ lævi, obliquè triangulari, crassâ, tumidâ, posticè subbiangulari, inæqnilaterali; valvulis percrassis, anticè crassioribus; natibus valdè elevatis, crassis; epidermidè tenebroso-fuscâ, posticè luteolâ, eradiatâ: dentibus cardinalibus subgrandibus, erectis, compressis, corrugatis, in utroque valvulo duplicibus; lateralibus brevibus, percrassis, obliquis corrugatisque: margaritâ argenteâ et iridescente.

Hab.—Coosa River, Watnmpka, Alab., E. R. Sbowalter, M. D.

Unio dolosus.—Testâ lævi, obovatâ, subalatâ, subcompressâ, ad laterè planulatâ, posticè et anticè rotundatâ, valdè inæquilaterali; valvulis subtenuibus: natibus prominulis, ad apices minutè undulatis; epidermide virido-olivâ et obsoletè radiatâ; dentibus cardinalibus parvis, compressis crenulatisque: lateralibus longis, lamellatis subcurvisque; margaritâ albidâ et purpureâ, paulisper tinctâ et valdè iridescente.

Hab.—Alabama River, Claiborne, Alab., Judge Tait. Coosa River, E. R. Showalter, M. D.

Unio parvulus.—Testâ plicatâ, transversâ, subinflatâ, ad laterè compressâ. posticè obtusè angulatâ, valdè inæquilaterali; valvulis tenuibus, anticè paulisper crassioribus; natibus prominulis; epidermidè olivaceâ, subvirido-maculatâ; dentibus cardinalibus parvis, compressis, crenulatis, in utroque valvulo duplicibus; lateralibus longis subrectisque; margaritâ cæruleâ et iridescente
Hab.—Coosa River, Alab., E. R. Showalter, M. D.

Unio plancus.—Testâ lævi, obliquo-ovatâ, subcompressâ, posticè biangulatâ anticè rotundatâ, valdè inæquilaterali; valvulis crassiusculis, anticè paulisper crassioribus; natibus prominulis; epidermide luteo-fuscâ, radiatâ; dentibus cardinalibus parvis, erectis, crenulatis, in utroque valvulo duplicibus; lateralibus sublongis subrectisque; margaritâ cæruleo-albâ et valdè iridescente.
Hab.—Coosa River, at Watumpka, Alab., E. R. Showalter, M. D.

Anodonta Showalterii.—Testâ lævi, ellipticâ, ventricosâ, posticè obtusè angulatâ, anticè rotundatâ, subæquilaterali; valvulis crassiusculis, anticè paulisper crassioribus; natibus subprominentibus, ad apices minutè undulatis: epidermide tenebroso-fuscâ, obsoletè radiatâ; margaritâ vel albâ vel paulisper salmonis colore tinctâ et iridescente.
Hab.—Coosa River, Watumpka, Alab., E. R. Showalter, M. D.

Descriptions of Three New Species of Exotic UNIONIDÆ.

BY ISAAC LEA.

[Read July 3d, 1860.]

Unio occatus.—Testâ plicatâ, ellipticâ, rugoso-occatâ, compressâ, valdè inæquilaterali, posticè biangulatâ, anticè regulariter rotundatâ; natibus prominulis, valdè compressis, ad apices plicis pulchris divaricatis; epidermidè luteo-olivâ et valdè rugosâ; dentibus cardinalibus parvis, compressis, obliquis; lateralibus sublongis subcurvisque; margaritâ albâ et valdè iridiscente.
Hab.—Bengal, W. A. Haines.

Anodonta Cailliaudii.—Testâ lævi, rotundâ, ventricosâ, inæquilaterali, an-

21

ticè supernè angulatâ; valvulis crassis, anticè paulisper crassioribus; natibus elevatis, tumidis, incurvis; epidermide nigro-fuscâ, obsoletè radiatâ, supernè micante, infernè exilissimè striatâ; margaritâ argenteâ et valdè iridescente.

Hab.—Brazil, Monr. F. Cailliaud.

MYCETOPUS EMARGINATUS.—Testâ lævi, transversissimè, compressâ, emarginatâ, ad latere planulatâ, anticè inflatâ, posticè ampliatâ et compressâ; valvulis pertenuibus, diaphinis, natibus parvis, prominulis; epidermide luteo-corneâ. valdè striatâ, nitidâ, eraditâ; margaritâ cæruleo-albâ et valdè iridescente.

Hab.—Siam, S. R. Honse, M. D.

Descriptions of Twenty-five New Species of UNIONIDÆ from Georgia. Alabama, Mississippi, Tennessee and Florida.

BY ISAAC LEA.

(Read February 5th, 1861.)

UNIO FABACEUS.—Testâ lævi, oblongâ, subquadratâ, subinflatâ, posticè subbiangulatâ, subæquilaterali; valvulis crassiusculis, anticè crassioribus; natibus prominulis; epidermide tenebroso-fuscâ, micante, obsoletè radiatâ; dentibus cardinalibus parvis, erectis, acuminatis, crenulatis; lateralibus curtis, lamellatis subcurvisque: margaritâ purpurascente, salmonis colore tinctâ et valdè iridescente.

Hab.—Oostananla River, Georgia. Bishop Elliott.

UNIO IRRASUS.—Testâ lævi, rotundo-trigonâ, inflatâ, posticè obtusè angulatâ anticè rotundâ; valvulis subcrassis, anticè crassioribus; natibus subelevatis. crassis; epidermide luteo-fuscâ, vel obsoletè radiatâ vel eradiatâ; dentibus cardinalibus crassis, elevatis, subcompressis crenulatisque: lateralibus curtis. crassis, obliquis rectisque; margaritâ argenteâ et iridescente.

Hab.—Etowah River, Georgia. Rev. G. White.

UNIO OCMULGÉENSIS.—Testâ lævi, transversâ, inflatâ, posticè obtusè biangulatâ, anticè subtruncatâ, valdè inæquilaterali: valvulis crassis, anticè crassioribus; natibus prominulis; epidermide tenebroso-fuscâ, eradiatâ, supernè micante, infernè valdè striatâ; dentibus cardinalibus parvinsculis, pyramidatis striatisque; lateralibus prælongis, lamellatis subrectisque; margaritâ argenteâ et iridescente.

Hab.—Little Ocmulgee River, Lumber City, Georgia. S. W. Wilson, M. D.

UNIO CICUR.—Testâ lævi, oblongâ, subinflatâ, ad latere subplanulatâ, posticè rotundatâ, valdè inæquilaterali; valvulis tenuibus, subdiaphanis; natibus subprominentibus, ad apices undulatis; epidermide olivaceâ, eradiatâ: dentibus cardinalibus parvissimis, compressis subrectisque ; lateralibus longis, prætenuis, lamellatis subrectisque ; margaritâ cærnleâ et valdè iridescente.

Hab.—Little Ocmulgee River, Georgia. S. W. Wilson, M. D.

UNIO CRAPULUS.—Testâ lævi, obliquâ, ventricosâ, ad umbones valdè tumidâ, valdè inæquilaterali, posticè rotundatâ, anticè truncatâ; valvulis percrassis, anticè crassioribus; natibus valdè prominentibus crassisque; epidermide luteofuscâ, eradiatâ; dentibus cardinalibus percrassis, pyramidatis, corrugatis, in utroque valvulo duplicibus: lateralibus percrassis, corrugatis, obliquis subcurvisque : margaritâ albâ et paulisper iridescente.

Hab.—Etowah River, Georgia. Rev. G. White.

UNIO BEADLEIANUS.—Testâ lævi, subrotundâ, ventricosâ, suhæquilaterali, anticè rotundatâ, posticè obtusè angulatâ; valvulis crassis, anticè crassioribus. natibus subelevatis, incurvis; epidermide tenebroso-fuscâ, obsoletè radiatâ : dentibus cardinalibus magnis, erectis, compressis corrugatisque ; lateralibus crassis, curtis corrugatisque; margaritâ vel albâ vel roseâ, et iridescente.

Hab.—Pearl River, Jackson, Mississippi. Rev. E. R. Beadle.

UNIO CHICKASAWHENSIS.—Testâ lævi, subrotundâ, subcompressâ, sublenticulari, inæquilaterali, posticè obtusè angulatâ, anticè rotundâ ; valvulis crassiusculis, anticè paulisper crassioribus; natibus prominulis; epidermide tenebrosofuscâ, eradiatâ, excillissimè striatâ ; dentibus cardinalibus parviusculis, pyramidatis, corrugatis crenulatisque ; lateralibus brevibus, subvalidis subcurvisque ; margaritâ rosaceâ et valdè iridescente.

Hab.—Chickasawha River, Mississippi. W. Spillman, M. D.

UNIO CINNAMOMICUS.—Testâ lævi, ellipticâ, inflatâ, ad umbones tumidâ, inæquilaterali, posticè angulatâ, anticè rotundâ ; valvulis subcrassis, anticè crassioribus ; natibus subprominentibus ; epidermide cinnamomicâ, infernè striatâ, eradiatâ, dentibus cardinalibus parviusculis, erectis, subcompressis crenulatisque, lateralibus curtis subrectisque : margaritâ albidâ et valdè iridescente.

Hab.—Tombigbee River, Columbus, Mississippi. W. Spillman, M. D.

UNIO PAUPERCULUS.—Testâ lævi, subrotundâ, subcompressâ, subequilaterali, posticè subrotundâ, anticè rotundâ ; valvulis subcrassis, anticè crassioribus : natibus prominulis; epidermide luteo-corueâ, eradiatâ; dentibus cardinalibus magnis, elevatis, decussatis ; lateralibus brevissimis, obliquis rectisque ; margaritâ albâ et iridescente.

Hab.—Stream near Columbus, Mississippi. W. Spillman, M. D.

UNIO SPILLMANII.—Testâ lævi, ellipticâ, subinflatâ, inæquilaterali, posticè obtusè angulatâ, anticè rotundatâ; valvulis subcrassis, anticè paulisper crassiori-

bus; natibus prominulis; epidermide tenebroso-fuscâ vel luteo-fuscâ, ad umbones nitidâ, radiatâ; dentibus cardinalibus crassiusculis, obtusè pyramidatis,
corrugatis ; lateralibus longis, crassis corrugatisque ; margaritâ vel albâ vel
salmonis colore tinctâ et valdè iridescente.

Hab.—Luxpalila Creek, near Columbus, Mississippi.   W. Spillman, M. D.

Unio flavidulus.—Testâ lævi, ellipticâ, subinflatâ, valdè inæquilaterali, posticè obtusè angulatâ, anticè rotundâ ; valvulis subtenuibus, anticè crassioribus ;
natibus prominulis : epidermide vel luteo-fuscâ vel luteo-viridi, eradiatâ ; dentibus cardinalibus parviusculis, erectis, compressis, in utroque valvulo duplicibus ; lateralibus sublongis, lamellatis subrectisque ; margaritâ albâ et iridescente.

Hab.—Stream near Columbus, Mississippi.   W. Spillman, M. D.

Unio anaticulus.—Testâ lævi, obliquâ, ad umbones valdè tumidâ, anticè
truncatâ, posticè obtusè angulatâ, valdè inæquilaterali ; valvulis crassiusculis,
anticè crassioribus ; natibus elevatis, crassis, incurvis, ferè terminalibus ; epidermide castaneâ, vittatâ, obsoletè radiatâ, dentibus cardinalibus subcrassis,
subpyramidatis crenulatisque ; lateralibus crassis, obliquis subrectisque ; margaritâ argenteâ et iridescente.

Hab.—Near Columbus, Mississippi.   W. Spillman, M. D.

Unio rubidus.—Testâ sulcatâ, subtriangulari, valdè inflatâ, ad laterè planulatâ, subæquilaterali, valvulis subcrassis, anticè crassioribus; natibus subprominentibus, subinflatis; epidermide tenebroso-rufo-fuscâ, eradiatâ, supernè micante, infernè striatâ ; dentibus cardinalibus crassiusculis, elevatis, subpyramidatis crenulatisque ; lateralibus sublongis, curvis subcrassisque ; margaritâ
vel rosaceâ vel albâ vel salmonis colore tinctâ et iridescente.

Hab.—Tombigbee River, Mississippi. W. Spillman, M. D.; Coosa River and
Big Prairie Creek, Alabama.   E. R. Showalter, M. D.

Unio decumbens.—Testâ lævi, arcuatâ, valdè compressâ, ad laterè planulatâ,
inæquilaterali, posticè biangulatâ, anticè rotundâ ; valvulis subtenuibus, anticè
et posticè paulisper crassioribus ; natibus prominulis ; epidermide tenebroso-rufo-
fuscâ, obsoletè radiatâ, transversè striatâ ; dentibus cardinalibus minimis, subcompressis, in utroque valvulo duplicibus ; lateralibus prælongis, arcuatis ;
margaritâ purpurascente, et valdè iridescente.

Hab.—Alabama. E. R. Showalter, M. D.

Unio germanus.—Testâ lævi, ellipticâ, subinflatâ, inæquilaterali, posticè subbiangulatâ, anticè rotundâ; valvulis crassiusculis, anticè crassioribus ; natibus
subprominentibus, ad apices concentricè rugoso-undulatis ; epidermide tenebroso-fuscâ, eradiatâ, transversè striatâ ; dentibus cardinalibus parvis, erectis,

24

·ompressis, crenulatis, acuminatis; lateralibus tenuibus subcurvisque: margaritâ purpurascente et valdè iridescente.

Hab.—Coosa River, Alabama. E. R. Showalter, M. D.

Unio Lewisii.—Testâ lævi, subrotundâ, suborbiculari, subæquilaterali; valvulis crassissimis, anticè crassioribus; natibus elevatis, tumidis incurvisque: epidermide luteolâ, punctatâ; dentibus cardinalibus crassissimis, erectis crenulatisque; lateralibus crassissimis, brevibus et obliquis; margaritâ albâ et iridescente.

Hab.—Coosa River, Alabama. E. R. Sbowalter, M. D.

Unio medius.—Testâ lævi, obliquâ, valdè inflatâ, valdè inæquilaterali, posticè obtusè angulatâ, anticè obliquè rotundatâ; valvulis crassis, posticè crassioribus: natibus elevatis, tumidis; epidermide fuscâ, maculatâ, infernè striatâ, supernè micante; dentibus cardinalibus crassis, pyramidatis crenulatisque; lateralibus crassis, rectis brevibusque; margaritâ argenteâ et iridescente.

Hab.—Coosa River, Alabama. E. R. Showalter, M. D.

Unio concolor.—Testâ lævi, obliquè ellipticâ, subinflatâ, inæquilaterali, posticè subbiangulatâ, anticè rotundâ; valvulis subcrassis, anticè crassioribus; natibus subprominentibus; epidermide tenebroso-olivâ, eradiatâ, ad umbones nitidâ, infernè striatâ; dentibus cardinalibus crassiusculis, erectis, obtusè compressis; lateralibus sublongis, obliquis subrectisque; margaritâ albâ et iridescente.

Hab.—Big Prairie Creek, Alabama. E. R. Sbowalter, M. D.

Unio venus.—Testâ lævi, subtriangulari, subcompressâ, valdè inæquilaterali, posticè ferè rotundâ, anticè rotundâ; valvulis crassiusculis, anticè crassioribus; natibus elevatis: epidermide tenebroso-olivâ, eradiatâ, maculatâ, vittatâ, ad umbones micante, infernè striatâ; dentibus cardinalibus parviusculis, compresso-pyramidatis striatisque; lateralibus, subbrevibus, obliquis subrectisque; margarita albâ et iridescente.

Hab.—Cabawba River, Perry Co., Alabama. E. R. Sbowalter, M. D.

Unio asperatus.—Testâ valdè tuberculatâ, subrotundâ, inflatâ, anticè et posticè rotundâ, subequilaterali; valvulis crassis, anticè crassioribus; natibus valdè prominentibus; epidermide rufo-luteâ, eradiatâ; dentibus cardinalibus percrassis, obtuso-conicis, corrugatis; lateralibus brevissimis, valdè obliquis rectisque; margaritâ argenteâ et iridescente.

Hab.—Alabama River, Claiborne, Alabama. Judge Tait.

Unio ornatus.—Testâ lævi, subrotundâ, compressâ, inæquilaterali, posticè subrotundâ, anticè rotundâ; valvulis crassiusculis, anticè crassioribus; natibus subprominentibus, ad apices rugoso-undulatis; epidermide melleâ, viridi maculatâ, supernè nitidâ, infernè striatâ; dentibus cardinalibus parviusculis, sulcatis; lateralibus brevibus, obliquis rectisque; margaritâ argenteâ et valdè iridescente.

Hab.—Alabama? T. R. Ingalls, M. D.

Unio peapurpureus,—Testâ lævi, ellipticâ, subinflatâ, inæquilaterali, posticè et anticè rotundatâ; valvulis subcrassis, anticè crassioribus; natibus prominulis; epidermide tenebroso-viridi, nigricente, radiis capillaris; dentibus cardinalibus parvinsculis, erectis, conicis, in utroque valvulo duplicibus, striatis; lateralibus longis rectisque; margaritâ valdè purpureâ et iridescente.

Hab.—Tennessee. J. G. Anthony.

Unio Anthonyi.—Testâ lævi, ellipticâ, inflatâ, ad latcrè planiusculâ, posticè obtusè biangulatâ, posticè rotundatâ, inæquilaterali; valvulis subtenuibus, anticè paulisper crassioribus; natibus prominulis; epidermide luteo-olivâ, eradiatâ; dentibus cardinalibus parvis, obliquis, subcompressis crenulatisque; lateralibus longis, lamellatis subcurvisque; margaritâ cæruleo-albâ et iridescente.

Hab.—Florida. J. G. Anthony.

Margaritana quadrata.—Testâ lævi, oblongâ, subcompressâ, ad laterè planulatâ, subæquilaterali, posticè obtusè angulatâ, anticè rotundâ; valvulis subtenuibus, anticè paulisper crassioribus; natibus prominulis, ad apices undulatis; epidermide luteolâ, viridi-radiatâ; dentibus cardinalibus subgrandibus, obliquis, compressis, triangularis, erectis subcurvisque; margaritâ albâ, supernè salmonis colore tinctâ, et valdè iridescente.

Hab.—East Tennessee, President Estabrook.

Margaritana Alabamensis.—Testâ lævi, oblongâ, inflatâ, ad laterè paulisper planulatâ, inæquilaterali, posticè obtusè biangulatâ, anticè obliqne rotundatâ; valvulis subcrassis, anticè paulisper crassioribus; natibus prominulis, ad apices rugoso undulatis; epidermide luteo-olivâ, politâ, eradiatâ; dentibus cardinalibus parvis, suberectis; margaritâ albâ et salmoniâ et iridescente.

Hab.—Talladega Creek, Alabama. W. Spillman, M. D.

4

Description of a new species of NERITINA, from Coosa River, Alabama.

BY ISAAC LEA.

(Read February 12, 1861.)

P. 35-78 NERITINA SHOWALTERII.—Testâ lævi, rotundatâ, diaphanâ, luteo-corneâ; spirâ valdè depressâ; suturis leviter impressis; anfractibus trinis, inflatis; aperturâ semirotundâ; labio dilatato, albo, incrassato, edentulo et incurvato; labro dilatato, tenui, margine acuto.

OPERCULUM ———.

Hab.—Coosa River, ten miles above Fort William, Shelby Co., Alabama. E. R. Showalter, M. D.

Remarks.—The discovery of this shell by Dr. Showalter marks the first notice of the genus *Neritina* being observed in our fresh waters. His very close observation and active investigation of the waters of central and northern Alabama, have enabled him to lay the naturalists of this country under many obligations by new discoveries, and this is certainly one of much importance. We now see for the first time that this genus, which is common in Europe, Africa, Asia, South America and the West Indies, also inhabits our southern fresh waters. I have great pleasure in naming the species after the discoverer.

This species is not closely allied to any which has come under my notice. It is more rotund than usual, has a clear horn-colored epidermis, smooth and shining; the substance of the shell so thin as to permit the column to be visible through it.

It is to be regretted that among the four specimens sent to me by Dr. Showalter, neither had an operculum. The soft parts of the animal have not yet been observed.

Descriptions of two new species of ANODONTÆ, from Arctic America.

BY ISAAC LEA.

(Read February 12, 1861.)

ANODONTA KENNICOTTII.—Testâ lævi, ellipticâ subinflatâ, inæquilaterali, posticè obtusè biangulatâ, anticè rotundâ; valvulis subtenuibus; natibus prominentibus, acuminatis, ad apices granulatis; epidermide pallido-luteâ usque tenebroso-fuscâ, eradiatâ; margaritâ cæruleo-albâ et iridescente.

Hab.—Great Slave Lake at Fort Rae, and north end of Lake Winnipeg, Arctic America. R. Kennicott.

ANODONTA SIMPSONIANA.—Testâ lævi, ellipticâ, subcompressâ, elongato-lenticulari, posticè obtusè angulatâ, anticè rotundâ; valvulis tenuibus; natibus prominulis, ad apices undulatis; epidermide tenebroso-fuscâ, eradiatâ; margaritâ cæruleo-albâ et iridescente.

Hab.—Fort Rae, Great Slave Lake, Arctic America. R. Kennicott.

Descriptions of New Species of SCHIZOSTOMA, ANCULOSA and LITHASIA.

BY ISAAC LEA.

(Read February 19th, 1861.)

SCHIZOSTOMA SPILLMANII.—Testâ striatâ, subcylindraceâ, subcrassâ, luteo-fuscâ, imperforatâ; spirâ obtnsâ, conoideâ; suturis impressis; anfractibus senis, valdè vittatis, planulatis, ultimo grandi; fissurâ obliquâ subbrevique; aperturâ grandi, ovatâ, intus vittatâ, ad basim obtusè angulatâ; columellâ albâ, supernè incrassatâ; labro-acuto sinuatoque.

Hab.—Coosa River, Alabama. Dr. Showalter.

ANCULOSA TURBINATA.—Testâ lævi, subrotundâ, crassâ, ponderosâ, tenebrosocorneâ, trivittatâ; spirâ obtusâ, vix exertâ; suturis valdè impressis; anfractibus quaternis. ultimo pergrandi; aperturâ magnâ, ovatâ, intus albidâ, trivittatâ, ad basim recurvatâ; columellâ incurvâ, impressâ; labro acuto, expanso, sinuato.

Hab.—North Alabama. Prof. M. Tuomey and Dr. Lewis. Tuscaloosa, Dr. Bndd.

ANCULOSA LEWISII.—Testâ lævi, ellipticâ, subcrassâ, subinflatâ, luteo-corneâ; spinâ obtusâ, vix exertâ, acuminatâ; suturis vix impressis; anfractibus quinis, ultimo pergrandi; aperturâ magnâ, regulariter ovatâ, intus albidâ; collumellâ incurvâ, supernè et infernè paulisper incrassatâ; labro acuto, subexpanso, paulisper sinuato.

Hab.—Tennessee. James Lewis, M. D.

Anculosa Coosaensis.—Testâ lævi, obtuso-conicâ, crassâ, tenebroso-corneâ, valdè vittatâ ; spirâ exertâ, ad apices obtusâ ; suturis valdè impressis ; anfractibus quaternis, infernè suturis valdè constrictis, ultimo magno ; aperturâ rotundatâ, albidâ, intus valdè vittatâ ; columellâ incrassatâ, incurvâ, tenebroso-purpureâ ; labro acuto, expanso.

Hab.—Coosa River, Alabama.   E. R. Showalter, M. D.

Lithasia fusiformis.—Testâ sulcatâ, fusiformi, subtenui, rufo-fuscâ, quadro-vittatâ ; spirâ conoideâ ; suturis irregulariter impressis ; anfractibus senis, ultimo magno et paulisper inflato ; aperturâ elongato-rbomboideâ, intus albidâ, quadro-vittatâ, ad basim canaliculatâ et recurvatâ ; columellâ sigmoideâ, supernè incrassatâ ; labro subconstricto, margine acuto.

Hab.—Coosa River, Alabama.   E. R. Showalter, M. D.

Lithasia imperialis.—Testâ tuberculatâ, fusiformi, subcrassâ, tenebroso-corneâ ; spirâ elevatâ, conoideâ ; suturis irregulariter et valdè impressis ; anfractibus senis, ultimo subgrandi, supernè irregulariter tuberculatis, subinflatis ; aperturâ parviusculâ, elongato-rbomboideâ, intus albidâ, fuscis capillaris instructis, ad basim canaliculatâ et recurvatâ ; columellâ sigmoideâ, supernè paulisper incrassatâ ; labro subexpanso, margine acuto.

Hab.—North Alabama.   Prof. Tuomey.

Lithasia Tuomeyi.—Testâ tuberculatâ, valdè inflatâ, subcrassâ, tenebroso-corneâ ; spirâ obtuso-conoideâ ; suturis impressis ; anfractibus quinis, ultimo grandi, infrà suturis obliquè tuberculatis ; aperturâ magnâ, rbomboideâ, intus albidâ, obsoletè vittatâ, ad basim canaliculatâ ; columellâ valdè incurvatâ, supernè et infernè incrassatâ; labro expanso, margine acuto.

Hab.—North Alabama.   Prof. M. Tuomey.

Lithasia subglobosa.—Testâ tuberculatâ, subglobosâ, crassâ, luteo-corneâ, bivittatâ ; spirâ vix exertâ; suturis impressis ; anfractibus quinis, ultimo grandissimo, apud humeris tuberculatis ; aperturâ magnâ, rbomboideâ, intus albâ, bivittatâ, ad basim canaliculatâ ; columellâ valdè incurvatâ, supernè et infernè valdè incrassatâ ; labro expanso, margine acuto.

Hab.—Tennessee.   Prof. G. Troost.

Lithasia dilatata.—Testâ lævi, subglobosâ, subcrassâ, glauco-virentè, infrà suturis luteolâ, obsoletè vittatâ ; spirâ obtusè conoideâ; suturis irregulariter impressis ; anfractibus quinis, ultimo magno et ventricoso ; aperturâ grandi, subrhomboideâ, intus fuscescente, ad basim angulatâ ; columellâ infernè et supernè incrassatâ, incurvâ; labro acuto et valdè dilatatâ.

Hab.—Tennessee.   Prof. G. Troost.

Descriptions of twelve new species of UNIONES, from Alabama.

BY ISAAC LEA.

(Read March 5th, 1861.)

UNIO NEGATUS.—Testâ sulcatâ, subtriangulari, compressâ, ad latere planulatâ, posticè obtusè angulatâ, anticè rotundâ, subæquilaterali ; valvulis subcrassis, anticè crassioribus ; natibus subprominentibus, acuminatis, ad apices corrugatis ; epidermide rufo-fuscâ, obsoletè radiatâ ; dentibus cardinalibus subgrandibus, striatis crenulatisque ; lateralibus subcrassis, sublongis subrectisque ; margaritâ vel albâ vel rosaceâ et iridescente.

Hab—Big Prairie Creek, Alabama. E. R. Showalter, M. D. And Columbus. Mississippi. W. Spillman, M. D.

UNIO GLANDACEUS.—Testâ lævi, subtriangulari, inflatâ, inæquilaterali, posticè subtriangulatâ, anticè rotundatâ ; valvulis crassis, anticè crassioribus ; natibus prominulis, crassis ; epidermide glandaceâ, rugosâ, eradiatâ ; dentibus cardinalibus magnis, valdè sulcatis, erectis ; lateralibus curtis, crassis, corrugatis. obliquis subrectisqne ; margaritâ albâ et iridescente.

Hab.—Cabawba River, Alabama. E. R. Sbowalter, M. D.

UNIO INSTRUCTUS.—Testâ lævi, subtriangulari. subcompressâ, inæquilaterali, posticè subbiangulatâ, anticè rotundâ ; valvulis crassiusculis, anticè crassioribus : natibus prominentibus, ad apices rugoso-undulatis ; epidermide melleâ, exillissimè striatâ, eradiatâ ; dentibus cardinalibus parviusculis, striatis crenulatisque : lateralibus subcurtis, striatis, obliquis subrectisque ; margaritâ argenteâ et iridescente.

Hab.—Cabawba River, Alabama. E. R. Showalter, M. D.

UNIO TRINACRUS.—Testâ lævi, triangulari, ad umbones tumidâ, inæquilaterali, posticè angulatâ, anticè obliquè rotundatâ ; valvulis crassis, anticè et posticè crassioribus ; natibus prominentibus, tumidis ; epidermide fusco-virente, obsoletè radiatâ, striatâ, dentibus cardinalibus parviusculis, depressis striatisque : lateralibus subcurtis, percrassis, obliquis, corrugatis rectisque ; margaritâ argenteâ et iridescente.

Hab.—Coosa River, Alabama. E. R. Showalter, M. D.

UNIO STABILIS.—Testâ lævi, triangulari, valdè tumidâ, valdè inæquilaterali. posticè subbiangulari, anticè rotundatâ, valvulis percrassis, anticè crassioribus : natibus valdè prominentibus, tumidis, solidissimis, incurvis ; epidermide pallido-melleâ, eradiatâ, infernè striatâ ; dentibus cardinalibus crassiusculis, com-

pressis, erectis striatisque; lateralibus crassis, curtis, obliquis, rectis corrugatisque ; margaritâ albâ et iridescente.

Hab.—Coosa River, Alabama. E. R. Showalter, M. D.

UNIO CONSANGUINEUS.—Testâ lævi, valdè obliquâ, anticè tumidâ et truncatâ
posticè compressâ et obtusè angulatâ ; valvulis crassis, anticè paulisper crassioribus, natibus tumidis, elevatis, incurvis terminalibusque ; epidermide luteo-
castaneâ, obsoletè radiatâ, transversè vittatâ ; dentibus cardinalibus subgrandibus, striatis subcompressisque ; lateralibus longis, crassis, corrugatis subcurvisque ; margaritâ argenteâ et iridescente.

Hab.—Etowah River. Rev. G. White. Oostenaula River, Georgia. Bishop
Elliott. And Cahawba River, Alabama. E. R. Showalter, M. D.

UNIO CREBRIVITTATUS.—Testâ lævi, valdè obliquâ, anticè tumidâ et truncatâ,
posticè compressâ rotundâque ; valvulis crassis, anticè crassioribus ; natibus
tumidis, elevatis, incurvatis terminalibusque; epidermide tenebroso-fuscâ, transversè et crebrè vittatâ ; dentibus cardinalibus subgrandibus, striatis corrugatisque ; lateralibus longis, crassis, corrugatis subcurvisque ; margaritâ argenteâ et
iridescente.

Hab.—Coosawattee River, Alabama. Bishop Elliott.

UNIO INTERVENTUS.—Testâ lævi, subobliquâ, subcompressâ, inæquilaterali,
posticè rotundatâ, anticè rotundâ ; valvulis crassiusculis, anticè crassioribus ;
natibus elevatis ; epidermide luteo-corneâ, supernè radiatâ, infernè striatâ, ad
umbones micanti ; dentibus cardinalibus parvis, pyramidatis striatisque ; lateralibus subcurtis, crassis, subobliquis subcurvisque ; margaritâ argenteâ et
valdè iridescente.

Hab.—Cahawba River, Alabama. E. R. Showalter, M. D.

UNIO PALLIDOFULVUS.—Testâ lævi, obliquâ, tumidâ, valdè inæquilaterali,
posticè rotundatâ, antice rotundâ ; valvulis crassis, anticè crassioribus ; natibus
elevatis, subincurvis ; epidermide pallido-fulvâ, maculatâ, infernè striatâ ; dentibus cardinalibus parvis, pyramidatis striatisque ; lateralibus subcurtis, crassis,
subobliquis ; margaritâ argenteâ et iridescente.

Hab.—Cahawba River, Alabama. E. R. Showalter, M. D.

UNIO PORPHYREUS.—Testâ lævi, ellipticâ, ventricosâ, valdè inæquilaterali,
posticè obtusè biangulatâ, anticè rotundatâ ; valvulis subcrassis, anticè crassioribus; natibus prominulis ; epidermide rufo-fuscescente, micanti, eradiatâ :
dentibus cardinalibus crassiusculis, corrugatis, crenulatis, in utroque valvulo
duplicibus ; lateralibus longis, subcrassis, corrugatis subrectisque : margaritâ
saturate-purpureâ et valdè iridescente.

Hab.—Coosa River, Alabama. E. R. Showalter, M. D.

Unio perpastus.—Testâ lævi, ellipticâ, valdè ventricosâ, valdè inæquilaterali, posticè obtusè biangulatâ, anticè obliquè rotundatâ; valvulis crassiusculis, anticè paulisper crassioribus; natibus subprominentibus, inflatis; epidermide luteo-fuscescente, supernè micanti, infernè striatâ, eradiatâ; dentibus cardinalibus parviusculis, erectis, conicis corrugatisque; lateralibus sublongis, lamellatis, corrugatis subcurvisque; margaritâ albâ et iridescente.

Hab.—Coosa River, Alabama. E. R. Sbowalter, M. D.

Unio granulatis.—Testâ plicatâ, ellipticâ, subinflatâ, valdè inæquilaterali, posticè obtusè angulatâ, anticè rotundâ; valvulis subtenuibus, anticè paulisper crassioribus, natibus prominulis, ad apices undato-granulatis; epidermide tenebroso-olivâ, eradiatâ, transversè striatâ; dentibus cardinalibus parvis, compressis, obliquis, crenulatis, in utroque valvulo duplicibus; lateralibus longis, acicularis, tenuis subrectisque; margaritâ purpurascente et valdè iridescente.

Hab.—Big Prairie Creek, Alabama. E. R. Sbowalter, M. D.

Description of a new genus (STREPHOBASIS) of the family MELANIDÆ, and three new species.

BY ISAAC LEA.

(Read April 16th, 1861.)

Family MELANIDÆ.

Genus STREPHOBASIS.

Testa cylindracea. Apertura subquadrata. Columella infernè incrassata et retro-canaliculata. Operculum corneum, instar spiræ.

The mollusk for which I propose this genus, was sent to me by Wm. Spillman, M. D., of Columbus, Mississippi, and I have before me over a dozen specimens from a third to nearly an inch in length. The very great number of species of the genus *Melania*, makes it desirable to eliminate any group with characters sufficiently distinct to permanently recognize it. The very remarkable retrose callus at the base of the column, causing a lateral sinus, is characteristic of this genus.

Strephobasis Spillmanii.—Testâ lævi, cylindraceâ, crassiusculâ, vel tenebroso-fuscâ vel virente, valdè vittatâ, nitidâ; spirâ obtusâ, curtâ, ad apicem carinatâ; suturis irregulariter impressis; anfractibus supernè convexiusculis, ultimo constricto; aperturâ subgrandi, subquadratâ, intus cærulescenti et valdè vittatâ; labro acuto, sinuoso; columellâ sinuosâ, ad basim incrassatâ et retrò canaliculatâ.

Hab.—Tennessee River. Wm. Spillman, M. D.

Strephobasis cornea.—Testâ lævi, cylindraceâ, crassâ, corneâ ; spirâ obtusâ ; suturis irregulariter impressis ; anfractibus supernè convexiusculis, ultimo constricto ; aperturâ rhombo-quadratâ, intus luteo-albâ ; labro acuto, sinuoso ; columellâ sinuosâ, ad basim incrassatâ et retro-canaliculatâ.

Operculum small, ovate, spiral, dark brown, with the polar-point near the base.

Hab.—Tennessee River.   Wm. Spillman, M. D.

Strephobasis Clarkii.—Testâ lævi, cylindraceâ, subtenui, luteo-corneâ, trivittatâ ; spirâ valdè obtusâ, curtâ ; suturis irregulariter impressis ; anfractibus quinis, supernè convexiusculis, ultimo constricto ; aperturâ subgrandi, quadratâ, intus albidâ, valdè vittatâ ; labro acuto ; columellâ sinuosâ, ad basim albâ, incrassatâ et retro-canaliculatâ.

Hab.—Tennessee River, at Chattanooga, Tennessee.   Joseph Clark.

Descriptions of forty-nine New Species of the Genus MELANIA,

BY ISAAC LEA.

(Read May 21st, 1861.)

During the past and present years, I have read several papers describing new species of *Unionidæ* and *Melanidæ*, kindly sent to me by E. R. Showalter, M. D., of Uniontown, Alabama, a Correspondent of our Academy, who has been unremitting in his exertions to make known the natural history of that part of the State. In these papers there were few species of the genus *Melania*. They were purposely delayed with a view to bring them as much together as possible; and the present paper will exhibit the vast expansion there of Zoological life in this single genus, the Coosa River really appearing to be the Zoological centre of this particular group.

The great variety of form, color and size will at once strike the Naturalist, and he will be surprised in the examination of these forms to observe how few there are of tuberculate or plicate species, which so well characterise the members of the same family, in the streams which form the Tennessee and Cumberland rivers at no great distance.

MELANIA HARTMANIANA.—Testâ lævi, conicâ, magnâ, vel tenebroso-corneâ vel tenebroso-oliva, valdè vittatâ, imperforatâ; spirâ obtusè conicâ; suturis valdè impressis; anfractibus subplanulatis, instar septenis, ultimo grandi; aperturâ grandi, ovato-rhomboideâ, intus brunneo-vittatâ, ad basim obtusè angulatâ; labro acuto; columellâ incurvatâ.

Hab.—Coosa and Cahawba Rivers, Alabama. E. R. Showalter, M. D.

MELANIA LEWISII.—Testâ striatâ, subcylindraceâ, tenebroso-virente, valdè vittatâ; spirâ subelevatâ, conoideâ; suturis valdè impressis; anfractibus planulatis, sulcatis, instar senis; aperturâ parviusculâ, ovato-rhomboideâ, intus valdè vittatâ, ad basim obtusè angulatâ; labro acuto; columellâ albâ et incurvatâ.

Hab.—Coosa and Talapoosa Rivers, Alabama. E. R. Showalter, M. D.

MELANIA ELLIPTICA.—Testâ lævi, ellipticâ, luteolâ, quadrivittatâ; spirâ brevi, obtusâ, ad apicem plicatâ; suturis impressis; anfractibus senis, subconvexis; aperturâ subgrandi, elongato-ellipticâ, intus quadrivittatâ, ad basim obtusè angulatâ; labro acuto; columellâ albidâ et incurvatâ.

Hab.—Coosa River, Alabama. E. R. Showalter, M. D. and E. Foreman, M. D.

MELANIA RUBICUNDA.—Testâ valdè striatâ, rubidâ, subfusiformi; spirâ subelevatâ, conoideâ; suturis impressis; anfractibus instar senis, convexiusculis;

5

34

aperturâ subconstrictâ, elongato-ellipticâ, intus rubidâ, ad basim obtuso-angulatâ : labro acuto ; columellâ incrassatâ, rubidâ, incurvatâ.

Hab.—Coosa River, Alabama. E. R. Sbowalter, M. D.

Melania vesicula.—Testâ lævi, ellipticâ, lutcâ, immaculatâ, subtcnui ; spirâ brevissimâ, obtusâ; suturis subimpressis ; anfractibus ternis, subconvexis ; aperturâ grandi, regulariter ovatâ, intus dilute-salmoniâ ; labro acuto ; columellâ incrassatâ, incurvatâ, ad basim rotuudatâ.

Hab.—Alabama. F. R. Showalter, M. D.

Melania Coosaensis.—Testâ striatâ, fusiformi, corneâ, quadrivittatâ, subcrassâ; spirâ subelevatâ, conicâ; suturis valdè impressis ; anfractibus septenis, convexiusculis, sulcatis; apertura constrictâ, elongato-ellipticâ, intus albidâ, et quadrivittatâ ; labro acuto, subcrenulato ; columellâ paulisper incrassatâ, incurvatâ, ad basim obtusè angulatâ.

Hab.—Coosa River, Alabama. F. R. Sbowalter, M. D.

Melania gracilior.—Testâ striatâ, fusiformi, viridi-lutescente, subcrassâ ; spirâ subelevatâ, conicâ; suturis irregulariter impressis ; anfractibus septenis, vix convexis ; apertura subconstrictâ, elongato-ellipticâ, intus albidâ ; labro acuto ; columellâ albidâ, infernè paulisper recurvâ, ad basim subrotundatâ.

Hab.—Coosa River, Alabama. E. R. Showalter, M. D.

Melania propria.—Testâ lævi, fusiformi, luteo-olivâ, quadrivittatâ, subcrassâ; spirâ obtuso-conoideâ ; suturis impressis ; anfractibus senis, convexiusculis ; apertura subgrandi, clongato-ellipticâ, intus albidâ et vittatâ ; labro acuto ; columellâ inflectâ, albâ, ad basim subangulatâ.

Hab.—Alabama. E R. Sbowalter, M. D.

Melania nubila.—Testâ striatâ, subellipticâ, obtusè couoideâ, tenebroso-vircute, obscurè maculatâ vel latè vittatâ, subcrassâ; spirâ obtusè elevatâ ; suturis irregulariter impressis ; anfractibus senis, subinflatis, ultimo grandi; apertura subgrandi, rhomboido-ellipticâ, intus quadrivittatâ ; labro acuto ; columellâ arcuatâ, ad basim obtusè angulatâ.

Hab.—Coosa River, Wetumpka, Alabama. F. R. Showalter, M. D.

Melania orbicula.—Testâ striatâ, globosâ, subcrassâ, luteo-virente, quadrivittatâ ; spirâ brevi, obtusâ ; suturis valdè impressis ; aufractibus quinis, valdè inflatis, ultimo graudi ; apertura grandi, ellipticâ, intus quadrivittatâ ; labro acuto ; columellâ albâ, incurvatâ, ad basim obtusè augulatâ.

Hab.—Coosa River, Alabama. E. R. Showalter, M. D.

Melania calculoides.—Testâ striatâ, subglobosâ, crassâ, corneâ, robustâ ; spirâ conicâ, valdè obtusâ ; suturis impressis ; anfractibus senis, valdè inflatis, ultimo grandi; aperturâ subgrandi, elongato-ellipticâ, iutus'albidâ ; labro acuto; columellâ albidâ, incrassatâ, arcuatâ, ad basim retusâ.

Hab.—Coosa River. Alabama. E. R. Showalter, M. D.

MELANIA PUNICEA.—Testâ lævi, subcylindraceâ, crassâ, puniceâ; spirâ elevatâ, conicâ; suturis impressis; anfractibus convexiusculis; aperturâ parvâ. rotundo-ovatâ, iutus albâ; labro acuto; columellâ incrassatâ, albâ, ad basim rotundatâ.

Hab.—Coosa River, Alabama. E. R. Showalter, M. D.

MELANIA LUTEOLA.—Testâ lævi, subellipticâ, subtenui, pallido-luteâ; spirâ subelevatâ, conoideâ; suturis paulisper impressis; anfractibus planiusculis; aperturâ subgrandi, iutus albidâ; labro acuto; columellâ albidâ, incurvâ; ad basim obtuso-angulatâ.

Hab.—Alabama River. E. R. Showalter, M. D.

MELANIA FASCINANS.—Testâ lævi, subfusiformi, crassiusculâ, luteo-corneâ, nitidâ; spirâ elevato-conicâ; suturis impressis; anfractibus convexiusculis; aperturâ subgrandi, intus albâ, trivittatâ; labro acuto; columellâ albâ, ad basim retusâ.

Hab.—Yellowleaf Creek, Shelby County, Alabama. E. R. Showalter, M. D.

MELANIA QUADRIVITTATA.—Testâ lævi, subellipticâ, crassiusculâ, viridi-luteâ, nitidâ; spirâ obtusè conoideâ; suturis valdè impressis; anfractibus octonis, convexiusculis; aperturâ subconstrictâ, rhombo-ovatâ, intus albidâ, quadri-vittatâ; labro acuto; columellâ incurvâ, ad basim angulatâ.

Hab.—Coosa River, Alabama. E. R. Showalter, M. D.

MELANIA MIDAS.—Testâ lævi, cylindraceo-ellipticâ, crassiusculâ, vireute, obsoletè vittatâ; suturis irregulariter impressis; anfractibus compressiusculis, ultimo pergrandi, infernè obsoletè striatâ; aperturâ grandi, auriculæformis, intus cæruleo-albâ; labro acuto; columellâ cæruleo-albâ, incrassatâ, inflectâ, ad basim obtusè angulatâ.

Hab.—Coosa and Alabama Rivers, near Wetumpka. E. R. Showalter, M. D.

MELANIA VARIATA.—Testâ lævi, subfusiformi, obtuso-conicâ, crassiusculâ. vel luteolâ vel purpurescente; suturis irregulariter impressis; anfractibus senis, supernè planinsculis, ultimo inflato; aperturâ grandi, intus vel luteolâ vel purpurescente; labro acuto; columellâ arcuatâ, inspissatâ, ad basim obtusè angulatâ.

Hab.—Coosa River, at Wetumpka and Montevallo, Bibb County, Alabama. E. R. Showalter, M. D.

MELANIA VIRGULATA.—Testâ lævi, fusiformi, conicâ, crassiusculâ, nitidâ, mucronatâ, luteolâ, quadrivittatâ ; suturis subimpressis ; anfractibus septenis, superné constrictâ, ultimo bulboso; aperturâ subgrandi, subellipticâ, intus luteo-albâ et valdé vittatâ ; labro acuto ; columellâ inflectâ, ad basim angulatâ et canalieulatâ.

Hab.—Coosa and Tallapoosa Rivers, Alabama. E. R. Showalter, M. D.

MELANIA MUCRONATA.—Testâ lævi, acuto-conoideâ, tenui, diaphanâ, stramineoluteâ; spirâ exertâ, mucronatâ; suturis leviter impressis ; anfractibus senis, superné planulatis : aperturâ parviusculâ, ovato-rhomboideâ, intus luteoalbidâ ; labro acuto, sinuato ; columellâ ad basim paulisper incrassatâ, subeffusâ et subrecurvâ.

Hab.—Big Prairie Creek, Alabama. E. R. Showalter, M. D.

MELANIA PROPINQUA.—Testâ lævi, subcylindraceâ, subcrassâ, luteolâ, quadrivittatâ ; spirâ subelevatâ, conoideâ ; snturis valdé impressis ; anfractibus senis, superné planiusculis; aperturâ ellipticâ, parviusculâ, intus albidâ et vittatâ; labro acuto ; columellâ paulisper incrassatâ, inferné rotundatâ.

Hab.—Coosa and Cahawba Rivers, Alabama. E. R. Showalter, M. D.

MELANIA SUAVIS.—Testâ lævi, subfusiformi, subcrassâ, luteo-viridi, politâ, quadrivittatâ ; spirâ obtuso-conica ; suturis regulariter impressis ; anfractibus senis, superné planiusculis ; aperturâ subgrandi, ellipticâ, iutus albidâ et vittatâ ; labro acuto ; columellâ incurvâ, ad basim rotundatâ.

Hab.—Coosa River, Alabama. E. R. Showalter, M. D.

MELANIA FALLAX—Testâ lævi, pupæformi, obtuso-conoideâ, subcrassâ, vel tenebroso-fuseâ vel tenebroso-corneâ, obsoleté vittatâ vel evittatâ ; suturis impressis ; anfractibus septenis, convexiusculis, ultimo parvo ; aperturâ parvâ, valdé constrictâ, elongato-ellipticâ ; labro acuto ; columellâ paulisper inflectâ, ad basim obtusé angulatâ.

Hab.—Coosa River, Alabama. E. R. Showalter, M. D.

MELANIA CLAUSA.—Testâ lævi, pupæformi, obtuso-conicâ, crassâ, olivâ, vittatâ vel evittatâ ; suturis valdé impressis ; anfractibus septenis, convexiusculis ; aperturâ parvâ, constrictâ, ellipticâ, iutus albidâ ; labro acuto ; columellâ paulisper inflectâ, ad basim obtusé angulatâ.

Hab.—Coosa River, Alabama. E. R. Showalter, M. D.

MELANIA PURPUREA.—Testâ lævi, subfusiformi, obtuso-conicâ, subtenui, tenebroso-rufâ ; suturis paulisper impressis : anfractibus quinis, ultimo grandi ; aperturâ subgrandi, ellipticâ, intus tenebrosâ ; labro acuto ; columellâ tenebrosâ, inflectâ.

Hab.—Alabama. E. R. Showalter, M. D.

MELANIA MELLEA.—Testâ lævi, subfusiformi, conica, crassiusculâ, mellcâ, aliquandò vittatâ; suturis irregulariter impressis: anfractibus septenis, supernè planulatis, ultimo grandi, inflato; aperturâ grandi, rhomboido-ellipticâ, intus luteolâ; labro acuto; columellâ incrassatâ, inflectâ, infernè obtusè angulatâ.

Hab.—Coosa River, at Wetumpka, Alabama.   E. R. Showalter, M. D.

MELANIA VARIANS.—Testâ lævi, vel plicatâ vel striatâ, elevato-conicâ, subcrassâ, luteolâ vel dilutè fuscâ, vittatâ; suturis impressis; anfractibus septenis, supernè planiusculis; aperturâ parviusculâ, ellipticâ, intus albidâ et vittatâ; labro acuto; columellâ albidâ, incurvatâ, ad basim obtusè angulatâ.

Hab.—Coosa River, Alabama.   E. R. Showalter, M. D.

MELANIA SHOWALTERII.—Testâ lævi, elevato-conicâ, subcrassâ, luteo-fuscâ, quadrivittatâ, suturis impressis; anfractibus instar senis, supernè planulatis, infernè subinflatis, ultimo subgrandi; aperturâ subgrandi, ovato-rhomboideâ, intus albidâ et vittatâ; labro acuto et paulisper sinuato; columellâ albâ, inflectâ, supernè paulisper incrassatâ, ad basim subrotundatâ.

Hab.—Coosa and Cahawba Rivers, Alabama.   E. R. Showalter, M. D.

MELANIA GLANDARIA.—Testâ lævi, obtuso-ellipticâ, crassâ, viridi-luteâ, quadrivittatâ; suturis valdè et irregulariter impressis; anfractibus septenis, convexiusculis, ultimo grandi; aperturâ elongato-ellipticâ, subconstrictâ, intus albâ et valdè vittatâ; labro acuto, subsinuoso; columellâ arcuatâ, supernè et infernè incrassatâ, paulisper canaliculatâ et contortâ.

Hab.—Coosa River, Alabama.   E. R. Showalter, M. D.

MELANIA PUDICA.—Testâ lævi, conoideâ, crassiusculâ, olivaceâ vel rufusculâ: suturis irregulariter impressis; anfractibus senis, convexiusculis; aperturâ parviusculâ, ovatâ, intus cæruleo-albâ; labro acuto; columellâ inflectâ, supernè incrassatâ, ad basim rotundatâ.

Hab.—Yellowleaf Creek, Alabama.   E. R. Showalter, M. D.

MELANIA SHELBYENSIS.—Testâ lævi, subellipticâ, subcrassâ, olivaceâ, vittatâ vel evittatâ: suturis impressis; anfractibus supernè planulatis; aperturâ parvinsculâ, subovatâ, intus albâ; labro acuto; columellâ inflectâ, ad basim obtusè angulatâ.

Hab.—Yellowleaf Creek, Alabama.   E. R. Showalter, M. D.

MELANIA ALABAMENSIS.—Testâ lævi, pupæformi, subelevatâ, subcrassâ, luteolâ, quadrivittatâ; suturis valdè impressis; anfractibus instar septenis, convexis: aperturâ parvâ, subconstrictâ, subellipticâ, intus albidâ et vittatâ; labro acuto; columellâ inflectâ, albidâ, ad basim obtusè angulatâ.

Hab.—Coosa River, Alabama.   E. R. Showalter, M. D.

Melania rara.—Testâ lævi, elevato-conoideâ, scalariformi, subcrassâ, tenebroso-olivâ, nitidâ; suturis irregulariter impressis; anfractibus octonis, planulatis, supernè angulatis; aperturâ parviusculâ, ellipticâ, intus tenebroso-purpureâ; labro acuto; columellâ incurvâ, purpureâ, ad basim obtusè angulatâ. *Hab.*—Coosa River, Alabama. E. R. Showalter, M. D.

Melania bullula.—Testâ lævi, conoideâ, inflatâ, subtenui, viridi-luteá, quadrivittatâ, suturis impressis; anfractibus instar quiuis, iuflatis, ultimo subgrandi; aperturâ subgrandi, latè ovatâ, intus albidâ et vittatâ; labro acuto; columellâ albidâ, supernè incrassatâ, sinuosâ, infernè subangulatâ. *Operculum* elliptical, spiral, dark brown, with polar point near the base. *Hab.*—Yellowleaf Creek, Shelby County, Alabama. E. R. Showalter, M. D.

Melania straminea.—Testâ lævi, regulariter ellipticâ, obtusè conoideâ, crassiuseulâ, stramineâ; suturis impressis; anfractibus quiuis, ultimo pergrandi et subinflatâ; aperturâ grandi, elongato-ellipticâ, intus luteo-albidâ; labro acuto; columellâ arcuatâ, supernè paulisper eallosâ, ad basim obtusè angulatâ. *Operculum* ovate, spiral, light brown, with the polar point near the edge towards the base. *Hab.*—Coosa River, Alabama. E. R. Showalter, M. D.

Melania solidula.—Testâ lævi, subfusiformi, obtusè eouicâ, erassiuseulá, luteo-viridi vel luteo-fuscâ, vittatâ; suturis impressis; anfraetibus quinis, supernè planulatis, infernè rotundatis, ultimo grandi; aperturâ subgrandi, ovatâ, intus albidâ; labro aeuto; columellâ arcuatâ, supernè paulisper callosâ, ad basim obtusè angulatâ. *Hab.*—Yellowleaf Creek, near its junetion with Coosa River, Alabama E. R. Showalter, M. D.

Melania Cahawbensis.—Testâ lævi, subfusiformi, elevato-conicâ, muoronatâ, subtenui, tenebroso-corncâ, obsoletè vittatâ; suturis linearibus; anfraetibus octonis, supernè planulatis, ultimo subgrandi; aperturâ parviusculâ, ovatâ intus albidâ vel luteolâ; labro acuto; columellâ arcuatâ, ad basim subrotundâ. *Hab.*—Cahawba River, Alabama. E. R. Showalter, M. D.

Melania culta.—Testâ rugoso-striatâ, obtuso-conoideâ, inflatâ, suberassá, viridi-luteâ, nitidâ, trivittatâ; suturis valdè et irregulariter impressis; anfractibus septenis, supernè carinatis; aperturâ amplâ, subrhomboideâ, intus albidâ et vittatâ; labro aeuto; eolumellâ incurvâ, dilutè roseâ, infernè angulatâ. *Hab.*—Coosa River, Alabama. E. R. Showalter, M. D.

Melania lita.—Testâ rugoso-striatâ, pupæformi, conoideâ, subcrassâ, quadrivittatâ, variegatâ, nitidâ; suturis irregulariter impressis; anfractibus senis, supernè convexis, ultimo elongato; aperturâ subconstrictâ, elongato-ovatâ, intus purpurescente et vittatâ; labro acuto, spissato; columellâ infernè incurvatâ, purpureâ, ad basim rotundatâ.

Hab.—Cahawba River, Alabama. E. R. Showalter, M. D.

Melania copiosa.—Testâ striatâ, latè ellipticâ, ventricosâ, obtuso-conicâ, crassiusculâ, luteo-corneâ, obsoletè vittatâ; suturis irregulariter impressis; anfractibus quinis, convexiusculis, ultimo pergrandi; aperturâ copiosâ, latè ellipticâ, intus albidâ; labro acuto, sinuoso; columellâ arcuatâ, supernè paulisper incrassatâ, ad basim subrotundâ.

Hab.—Coosa River, Alabama. E. R. Showalter, M. D.

Melania pergrata.—Testâ striatâ, subcylindraceâ, obtusè conicâ, crassiusculâ, viridi-corneâ, suturis valdè impressis; anfractibus senis, supernè humerosis, striis transversis crebrè indutis, ultimo pergrandi et cylindraceo; aperturâ grandi, elongato-ovatâ, intus albidâ; labro acuto; columellâ arcuatâ, supernè paulisper callosâ, ad basim subrotundatâ.

Operculum, ovate, spiral, dark brown, with the polar point on the edge near to the base.

Hab.—Coosa River, Alabama. E. R. Showalter, M. D.

Melania bellula.—Testâ striatâ, subellipticâ, obtusè conoideâ, crassiusculâ, luteo-corneâ, quadrivittatâ; suturis valdè impressis; anfractibus instar quinis, convexiusculis, ultimo grandi; aperturâ subgraudi, ellipticâ, intus albidâ et vittatâ; labro acuto; columellâ albâ, inflectâ, ad basim obtusè angulatâ.

Operculum elliptical, spiral, dark brown, with the polar point near the inner edge, about one-fourth from the base.

Hab.—Yellowleaf Creek, Shelby County, Alabama. E. R. Showalter, M. D.

Melania æqua.—Testâ substriatâ, conicâ, subcrassâ, tenebroso-fuscâ, suturis impressis; anfractibus instar senis, supernè planulatis; aperturâ parvâ, rhomboideâ, intus albidâ; labro acuto; columellâ inflectâ, paulisper incrassatâ, ad basim obtusè angulatâ.

Hab.—Yellowleaf Creek, Alabama. E. R. Showalter, M. D.

Melania capillaris.—Testâ crebrè striatâ, angustè ellipticâ, crassiusculâ, luteo-fuscâ, striis transversis capillaris crebressimè indutis; suturis irregulariter impressis; anfractibus subcompressis, ultimo grandi; aperturâ grandi, elongato-ellipticâ, intus striis capillaris; labro crenulato; columellâ albidâ, incrassatâ, incurvâ, ad basim obtusè angulatâ.

Operculum ovate, spiral, dark brown, with polar point near the inner side and near to the base.

Hab.—Coosa River, Alabama. E. R. Showalter, M. D. and Wm. Spillman, M. D.

MELANIA GRATIOSA.—Testâ tuberculatâ, aliquando striatâ, obtuso-fusiformi, crassiusculâ, luteo-viridi, vel vittatâ vel evittatâ ; suturis impressis ; anfractibus senis, superuè planulatis, ultimo grandi ; aperturâ subgrandi, subrhomboideâ, intus albidâ ; labro acuto, subsinuoso ; columellâ inflectâ, incrassatâ, ad basim subangulatâ.

Operculum ovate, spiral, dark brown, with the polar point near the base.
Hab.—Coosa River, Alabama. E. R. Showalter, M. D.

MELANIA PAULA.—Testâ carinatâ, conicâ, tenui, diaphanâ, rufo-corneâ ; suturis paulisper impressis ; anfractibus senis, supernè acuto-carinatis, ultimo sub-bicarinato ; aperturâ parviusculâ, lato-ellipticâ, intus albidâ ; labro acuto ; columellâ vel albidâ vel rufescente, inflectâ, ad basim acuto-angulatâ.

Hab.—Cahawba River, Alabama. E. R. Showalter, M. D.

MELANIA BLANDA.—Testâ plicatâ, obtusè fusiformi, supernè obtusè conicâ, subtenui, tenebroso-corneâ ; suturis impressis ; anfractibus quinis, supernè planulatis, ultimo grandi et subangulato ; aperturâ subgrandi, ellipticâ, intus luteo-albâ ; labro acuto ; columellâ incrassatâ, inflectâ, infernè subangulatâ.

Hab.—Yellowleaf Creek, Alabama. E. R. Showalter, M. D.

MELANIA CREPERA.—Testâ substriatâ, conicâ, subcrassâ, fuliginosâ ; spirâ subelevatâ ; suturis irregulariter impressis ; anfractibus senis, convexiusculis ; aperturâ ovato-rhombicâ, intus albidâ ; labro acuto ; columellâ inflectâ, supernè paulisper incrassatâ, ad basim obtusè angulatâ.

Hab.—Yellowleaf Creek, Shelby County, Alabama. E. R. Showalter, M. D.

MELANIA FUMEA.—Testâ lævi, conicâ, subtenui, fumeâ, subnitidâ, aliquando obsoletè vittatâ ; spirâ subelevatâ ; suturis irregulariter impressis ; anfractibus supernè planulatis, infernè subinflatis ; aperturâ ovato-rhombicâ, intus albidâ ; labro acuto ; columellâ inflectâ, supernè paulisper incrassatâ, ad basim subrotundâ.

Hab.—Yellowleaf Creek, Shelby County, Alabama. E. R. Showalter, M. D.

MELANIA PROPRIA.—Testâ lævi, elongato-ellipticâ, subtenui, luteo-corneâ, obsoletè vittatâ, nitidâ ; spirâ elevatâ ; suturis valdè impressis ; anfractibus instar senis, supernè convexiusculis, inferuè inflatis ; aperturâ subgrandi, ovatâ, intus luteo-albâ ; labro acuto ; columellâ inflectâ, supernè incrassatâ, ad basim rotundatâ.

Hab.—Yellowleaf Creek, Shelby County, Alabama. E. R. Showalter, M. D.

subcurtis subrectisque; margaritâ vel albâ vel roseâ vel salmoniâ et valdè iridescente.

Hab.—Dallas, Texas. Prof. C. G. Forshey.

Unio Heermannii.—Testâ alatâ, lævi, ellipticâ, compressâ, valdè inæquilaterali, posticè obtusè biangulatâ, anticè rotundâ; valvulis subtenuibus, anticè irregulariter crassioribus; natibus prominulis, vix undulatis; epidermide luteofuscâ, micanti, eradiatâ; dentibus cardinalibus parvis, subconicis, crenulatis, in utroque valvulo duplicibus; lateralibus longis, lamellatis subrectisque; margaritâ pallido-salmoniâ, purpurescente et intensè iridescente.

Hab.—Medina River, Texas. A. L. Heermann, M. D.

Unio Tesserclæ.—Testâ lævi, quadratâ, cuboideâ, valdè tumidâ, valdè inæquilaterali, posticè obtusè angulatâ, anticè truncatâ; valvulis crassis, ad apices rugoso-undulatis; epidermide melleâ, micanti, radiis interuptis indutis; dentibus cardinalibus parviusculis, subconicis corrugatisque; lateralibus curtis, obliquis rectisque; margaritâ argenteâ et valdè iridescente.

Hab.—Nolachucky River, Tenn. J. G. Anthony.

Unio Northamptonensis.—Testâ lævi, oblongâ, valdè compressâ, ad latere planulatâ, posticè obtusè biangulari, anticè obliquè rotundatâ, valdè inæquilaterali; valvulis subcrassis, anticè crassioribus; natibus prominulis; epidermide vel ochraceâ vel luteo-fuscâ, obliquè radiatâ; dentibus cardinalibus crassis, striatis, in utroque valvulo duplicibus; lateralibus prælongis, validis, corrugatis, subrectis lamellatisque; margaritâ vel albâ vel purpurascente vel salmonis colore tinctâ et valdè iridescente.

Hab.—Connecticut River, at Northampton. At Springfield, by L. Shurtleff, M. D. Below Hartford, T. R. Ingalls, M. D. Neuse River, N. C., E. Emmons, M. D.

Unio Wardii.—Testâ tuberculatâ, subtriangulari, compressâ, subæquilaterali, posticè et infernè emarginatâ, anticè rotundâ; valvulis crassiusculis, anticè crassioribus: natibus prominulis, ad apices rugosis; epidermide vel luteolâ vel luteo-virente, maculis triangularis indutis; dentibus cardinalibus subgrandibus, compressis sulcatisque; lateralibus sublongis, subcrassis, obliquis rectisque; margaritâ argenteâ, interdum roseâ et iridescente.

Hab.—Walbonding River, Ohio, J. C. Ward. Wassepinicon River, Iowa, Dr. Foreman. Coal River, Virginia, Dr. Hartman.

Unio Sampsonii.—Testâ lævi, oblongâ, inflatâ, ad umbones valdè tumidâ, posticè emarginatâ, anticè rotundâ, valdè inæquilaterali; valvulis crassis, anticè paulisper crassioribus; natibus prominentibus, tumidis, incurvis, ad apices vix undulatis; epidermide luteolâ, radiis viridis vestitis; dentibus cardinalibus

subgrandibus, erectis corrugatisque; lateralibus crassis, curtis, corrugatis sub·
rectisque; margarità argenteà et paulisper iridescente.

Hab.—Wabash River, New Harmony, Indiana. James Sampson.

UNIO VESTITUS.—Testâ lævi, ellipticâ, compressâ, inæquilaterali, posticè obtusè angulatâ, anticè rotundâ; valvulis subtenuibus, anticè paulisper crassioribus; natibus prominulis; epidermide vel luteâ vel lutco-fuscâ, politâ, radiis
obliquis viridis vestitis; dentibus cardinalibus parvis, compressis, acuminatis,
crenulatis, in utroque valvulo duplicibus; lateralibus sublongis, lamellatis,
subobliquis corrugatisque; margarità albidâ et splendidè iridescente.

Hab.—Ogechee River, Georgia. Major Le Conte and J. G. Anthony.

Descriptions of Seven New Species of the Genus IO.

BY ISAAC LEA.

Read December 24th, 1861,

When I proposed in 1831* to form the new genus *Io* for Mr. Say's *Fusus fluviatilis*, there were no other allied species known to naturalists. I then proposed also to change the specific name to *fusiformis*, as being more appropriate, and I gave a figure under this name. At that time the canons of nomenclature were not so well understood nor so strict as they have since been; and it is only justice to Mr. Say to relinquish my specific name, and to replace his. Subsequently in 1834, I proposed a new species under the name of *Io spinosa*, (Trans. Am. Phil. Soc., vol. v. pl. 19, fig. 79.) More recently Mr. Anthony, in the Proceedings of the Academy, (1860,) proposed four new species; three of which I think belong to the two previously established species. Mr. Lovell Reeve, in his beautiful "Conchologia Iconica," has recently issued among his monographs one of the genus *Io* with numerous plates and full descriptions. In this he has introduced a number of species, most of which I think more appropriately belong to Prof. Haldeman's genus *Lithasia*—the species of which form a very excellent group, which he separated from *Melania* and *Anculosa*—but which Mr. Reeve does not seem to recognise. Of the true *Io* I also think he has considered several varieties as species.

Io NONOSA.—Testâ tuberculatâ, elevato-conicâ, virido-corneâ, vittatâ; spirâ regulariter conicâ; suturis valdè impressis; anfractibus instar denis, planulatis, medio tuberculatis, infià striatis; aperturâ parviusculâ, rhomboideâ, intus vittatâ; labro acuto et sigmoideo; columellâ albâ et valdè contortâ; canali breviusculâ.

Hab.—Tennessee River, Alabama?† Wm. Spillman, M. D.

Io ROBUSTA.—Testâ canaliculatâ, paulisper tuberculatâ, elevato-conicâ, pallido-corneâ, infrà obsoletè vittatâ; spirâ regulariter conicâ; suturis valdè impressis; anfractibus instar denis, apud apicem planulatis, infrà canaliculatà: aperturâ parviusculâ, rhomboideâ, intus vittatâ; labro acuto et sigmoideo; columellâ pallido-salmoniâ; canali breviusculâ.

Hab.—Tennessee River, Alabama? Wm. Spillman, M. D.

---

* Trans. Amer. Phil. Soc., January, 1831.

† Dr. Spillman simply gave Tennessee River as the habitat of these species, but did not mention what part. They are probably from Alabama.

48

Io variabilis.—Testâ lævi, elevato-conoideâ, subfusiformi, vel vittatâ vel intensè purpureâ vel virente; spirâ regulariter conoideâ; suturis leviter impressis; anfractibus instar novenis, planulatis, in medio angulatis: aperturâ elongato-rhomboideâ; labro acuto et sinuoso; columellâ vel albidâ vel purpureâ et valdè contortâ; canali attenuato-constrictâ.

Hab.—Tennessee River, Alabama? Wm. Spillman, M. D.

Io modesta.—Testâ lævi, conicâ, virido-corneâ; spirâ regulariter conicâ; suturis impressis; anfractibus novenis, planulatis, in medio angulatis; aperturâ parvâ, regulariter rhomboideâ; labro acuto et sinuoso; columellâ albâ et valdè contortâ; canali curtâ et effusâ.

Hab.—Tennessee River, Alabama. Wm. Spillman, M. D.

Io Spillmanii.—Testâ lævi, attenuato-conicâ, pallido-corneâ; spirâ regulariter conicâ, supernè striatâ; suturis leviter impressis; anfractibus instar denis, planulatis, in medio obtusè angulatis; aperturâ parvâ, rhomboideâ; labro acuto et sinuoso; columellâ albâ et valdè contortâ; canali curtâ et subeffusâ.

Hab.—Tennessee River, Alabama? Wm. Spillman, M. D.

Io gracilis.—Testâ lævi, conicâ, pallido-purpureâ; spirâ regulariter conicâ: suturis regulariter impressis; anfractibus instar novenis, planulatis. in medio angulatis; aperturâ parviusculâ, rhomboideâ; labro acuto et sinuoso: columellâ pallido-purpureâ, valdè contortâ et deflectâ; canali curtâ et latè effusâ.

Hab.—Coosa River, Alabama. Wm. Spillman, M. D.

Io viridula.—Testâ lævi, cylindrico-conoideâ, virente; spirâ subelevatâ; suturis parum impressis; anfractibus instar novenis, planulatis, in medio obtusè angulatis; aperturâ parviusculâ, rhomboideâ; labro acuto, sinuoso; columellâ ad basim purpureâ, parum contortâ; canali curtâ et dilatatâ.

Hab.—Coosa River, Alabama. Wm. Spillman, M. D.

ERRATA.

Page 1, line 9, from bottom, for "Byssandonta," read Byssanodonta.

" 14, " 12, for "castanea," read castaneum.

" 14, " 24, for "globosa," read globosum.

" 15, " 5, for "Wetumpkaensis," read Wetumpkaense.

" 16, " 14, for "Coosa," read Cahawba.

" 16, " 14, for "Wetumpka," read Centreville.

" 34, " 15, "gracilior," changed to ellipsoides.

" 40, " 33, "propria" changed to lepida.

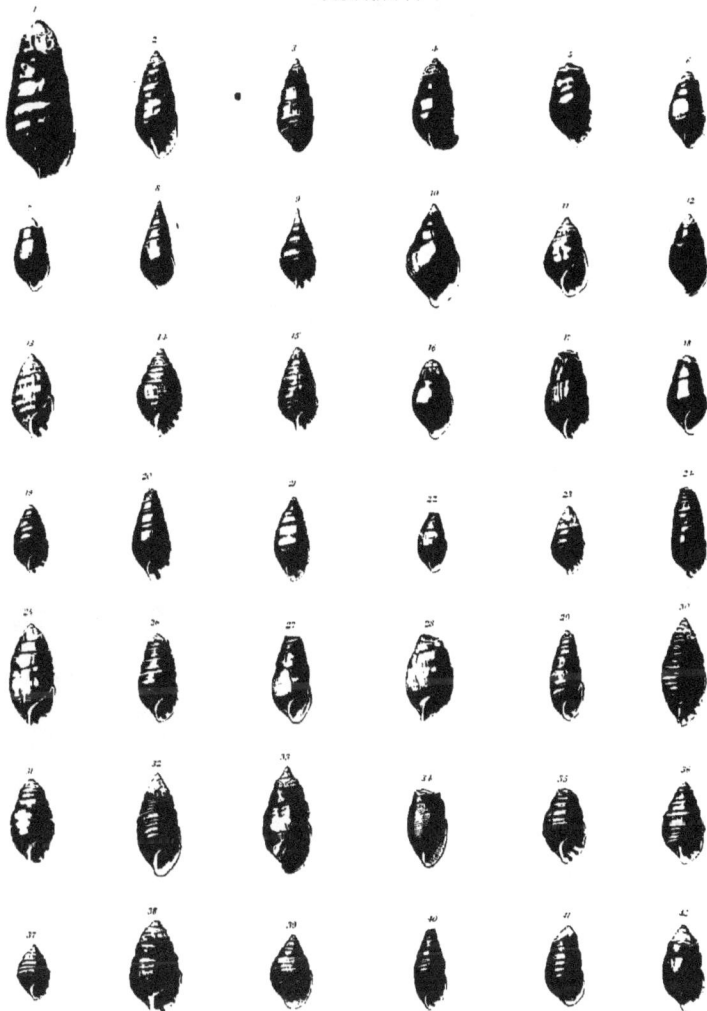

1	G. Hartmanni	9	G. virgulata	17	G. lepida	25	G. clausa	34	G. capillaris
2	G. varians	10	G. mellea	18	G. Shelbyensis	26	G. Alabamensis	35	G. bellula
3	G. rara	11	G. vinrata	19	G. suavis	27	G. pumeea	36	G. culta
4	G. Showalteri	12	G. purpurea	20	G. lacunosa	28	G. Midas	37	G. erbenia
5	G. bullula	13	G. elliptica	21	G. prospera	29	G. propinqua	38	G. calculoides
6	G. fumea	14	G. glandaria	22	G. luteola	30	G. Cassaensis	39	G. expansa
7	G. porluca	15	G. quadristriata	23	G. oblatula	31	G. ellipsoides	40	G. leta
8	G. Cahawbensis	16	G. straminea	24	G. Gillise	32	G. rubicunda	41	G. aqua
						33	G. nubila	42	G. prospera

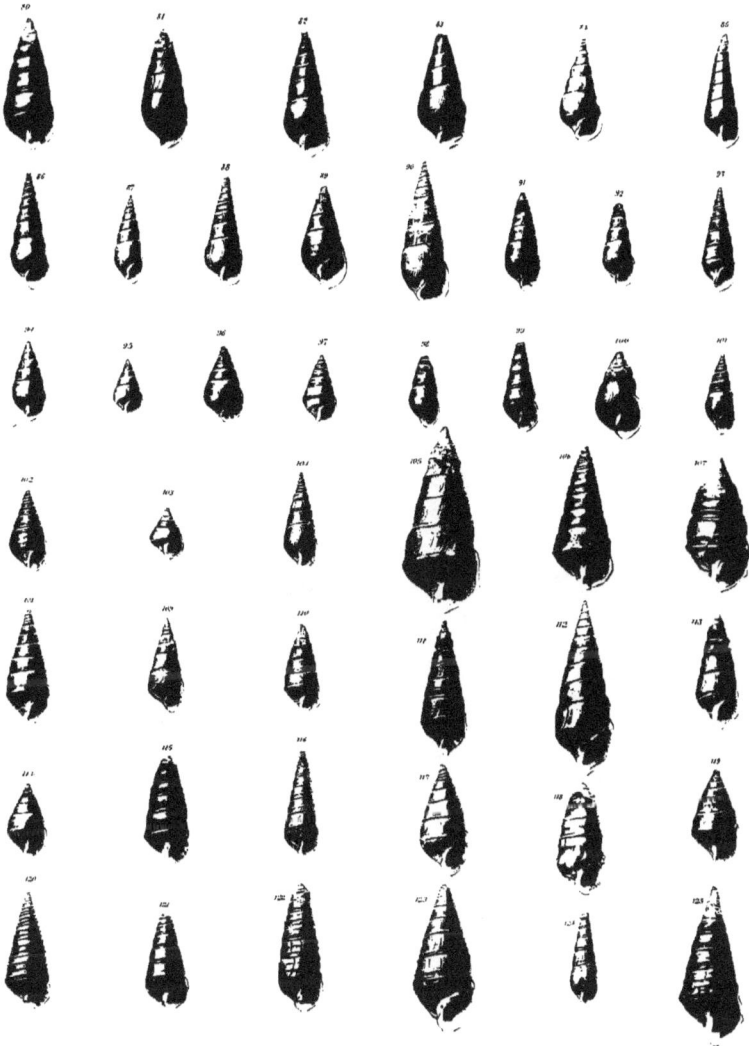

80	T	Hartmanni	89	T	tortum	98	T	Vanuxemi	107	V	Troosta	116	T	subulaforme
81	T	Joei	90	T	pallidum	99	T	chakasahense	108	V	Clarku	117	T	clracenum
82	T	Spillmani	91	T	parvum	100	T	Tennesseense	109	V	incurvum	118	T	noerforme
83	T	Chrestes	92	T	modestum	101	T	Knoxense	110	T	Patella	119	T	viride
84	T	labiatum	93	T	macromphalum	102	T	trevittatum	111	T	Thomeyi	120	T	Lewisii
85	T	Whitei	94	T	simplex	102	T	trochulus	112	F	Florencese	121	T	conahhum
86	T	Estabrooksii	95	T	imaor	103	T	dux	113	T	Alabamense	122	T	Showalteri
87	T	Knoxvillense	96	T	mundum	104	T	Thorntoni	114	T	ligatum	123	T	Sulkeru
88	T	attenuatum	97	T	brevitatum				115	T	Pybasii	124	T	gradium
												125	T	monditivum

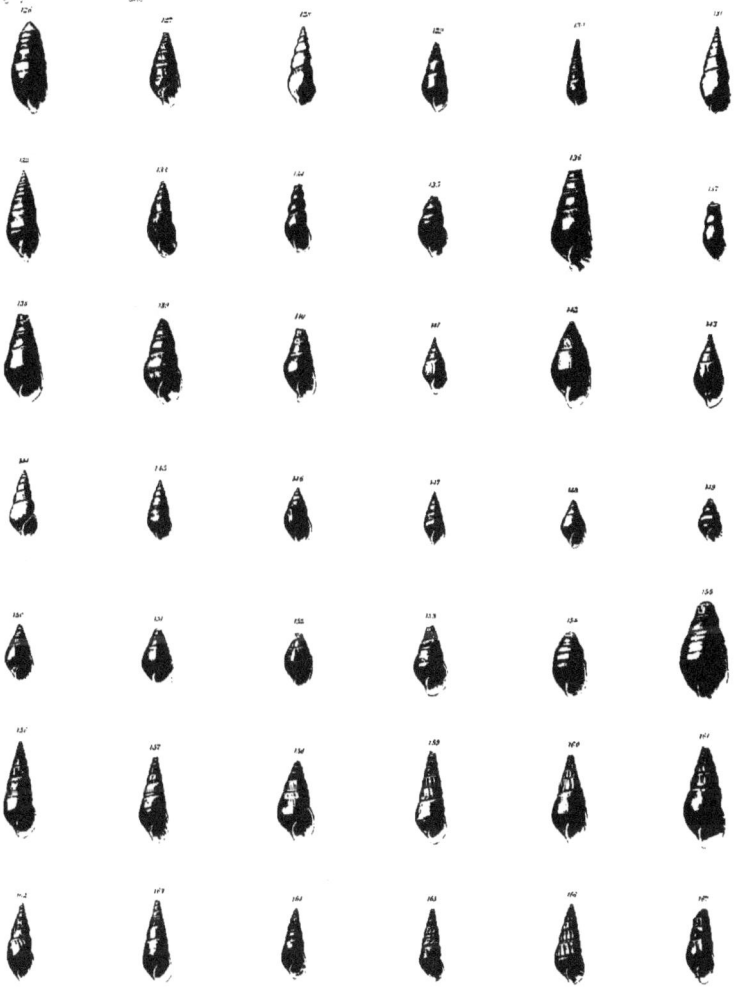

www.ingramcontent.com/pod-product-compliance
Lightning Source LLC
Chambersburg PA
CBHW022013190326
41519CB00010B/1510